A TIME TO GATHER STONES

A Time to Gather Stones

ESSAYS BY VLADIMIR SOLOUKHIN

translated with an introduction by
Valerie Z. Nollan

NORTHWESTERN UNIVERSITY PRESS
EVANSTON, ILLINOIS

Northwestern University Press
Evanston, Illinois 60208-4210

First published 1980 as *Vremya sobirat kamni* by Sovremennik,
Moscow. © 1980 by Vladimir Soloukhin. English translation
and introduction © 1993 by Northwestern University Press.
Published by arrangement with Vladimir Soloukhin.
All rights reserved.

Printed in the United States of America

Library of Congress Cataloging-in-Publication Data

Soloukhin, Vladimir Alekseevich.
   [Vremia sobirat' kamni. English]
   A time to gather stones : essays / by Vladimir Soloukhin ;
translated and with an introduction by Valerie Z. Nollan.
      p. cm.
   Includes bibliographical references.
   ISBN 0-8101-1127-6
   1. Cultural property, protection of—Soviet Union. 2. Historic
sites—Soviet Union—Conservation and restoration. 3. Literary
landmarks—Soviet Union. 4. Authors, Russian—Homes and haunts-
-Soviet Union. I. Nollan, Valerie Z. II. Title.
DK18.65.S6413   1993
363.6'9'0947—dc2   93-41119
                                            CIP

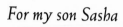

For my son Sasha

CONTENTS

PREFACE

I WAS BORN IN 1924 into a large peasant family, and we lived near the ancient Russian city of Vladimir. Hence I saw with my own eyes a large part of what took place in Russia during the course of entire decades, even if only from the example of our small hamlet, Olepino.

I remember collectivization and dekulakization, when peasant families were taken away to the nearest railroad station, and from there into the taiga and tundra to a certain—almost one-hundred-percent certain—death.

I remember how they threw down the bells from the bell towers. I remember the closing and destruction of the churches, and how ancient icons and books were thrown out of them. I remember how the militia took away first one person, then another, and how those people didn't return.

I saw all these things, just like the majority of my compatriots, but (just like the majority of my compatriots) I feel as if I didn't really see them. This was a delusion, temporary insanity, hypnosis.

Today our country, or perhaps the population of our country, has been awakened from its hypnotic spell, but nevertheless many people don't want to see everything that happened in our country. Vision is restored slowly and not in everyone.

I "recovered my sight" much earlier. The awakening of my social, civic thinking and national self-awareness dates to the very beginning of the 1960s. And what did I see when I opened my eyes? I saw that our land, our country, had been ruined, disfigured, and plundered. I learned that over ninety percent of our country's ancient monuments, in particular its cathedrals and monasteries, had been destroyed. I learned that in Moscow alone during the 1930s four hundred and fifty ancient cathedrals had been blown up. And not only cathedrals: the Red Gates, the Sukharev Tower, Kitai-gorod, Zaryadie, the monuments to General Skobelev, Alexander II, and Alexander III, Chudov Monastery and Voznesensky Monastery in the Kremlin, and Strastnoi Monastery, in whose location the movie theater Rossiya now stands, were also destroyed.

Throughout Russia country homes and estates with their marvelous buildings, ponds, and parks have been destroyed. It is impossible to write about all of them, impossible to mourn all of them. But I chose several sites, names, and locations, and went to see what remains today in the places where the poet Derzhavin, the poet Blok, and the bard of our own natural surroundings, Aksakov, had lived; where Optina Monastery, linked with the names of Dostoevsky, Tolstoi, and Gogol, had stood. . . .

The result of my travels to these places is the book *A Time to Gather Stones*. It appeared at a time when there weren't any perestroikas or shake-ups, and its publication was not a simple matter at all. It is enough to say that not a single copy of the first edition of 75,000 copies was put on sale; all of the copies were sent to the so-called library reserves. Anyone who knows the reality of those years understands that this was the most reliable method of hiding a book from its readers, from the people.

Time has passed and many things have changed, but our attitude toward our cultural, national sacred objects has not changed. To be sure, Optina Monastery has been returned to the Russian Orthodox Church, but concerning the others . . . Derzhavin's Zvanka is still in the same state of neglect, just as before. Stinging nettles still grow, just as before, in the place where Blok's Shakhmatovo had stood; the church in Tarakanovo, in which Blok was married to Lyubov Dmitrievna Mendeleeva, has finally collapsed; nothing has been done in Aksakovo. . . .

Will our grandchildren be kinder than we have been to our cultural treasures?

Vladimir Soloukhin
July 1993

ACKNOWLEDGMENTS

MY WORK ON this translation of Vladimir Soloukhin's essays over the last seven years has benefited enormously from the contributions of various individuals and institutions. I would especially like to thank Rhodes College for many forms of support: for awarding me two faculty development grants (for the summer months of 1988 and 1989); an early leave grant for the fall of 1989; a faculty travel grant for July 1991; and, finally, a sabbatical leave for the academic year 1992–93. All of these grants provided me with the time and material assistance necessary for any project of this scope.

Oberlin College and the Center for Russian and East European Studies at the University of Pittsburgh awarded me small travel grants that enabled me to interview Soloukhin in Moscow. My gratitude also goes to various individuals at the Synod of Bishops of the Russian Orthodox Church Outside of Russia in New York for researching some questions for me, and to Robert Burger and Richard Seitz of the Slavic Reference Service, University Library, University of Illinois, for their extensive reference work during the last stages of my translation. My research assistants at Rhodes College, Dennis Pannozza, Carla Carr Stec, and Vasily Shirpakov were exceptionally competent and helpful.

I remain indebted to several specialists for different reasons. Ludmila Koehler gave generously of her time in reading several drafts of three of the essays. Gerald Mikkelson and Kathleen Parthé provided useful comments on my work on Soloukhin at professional conferences over the years, while John Dunlop's impeccable scholarly work on Russian nationalism has informed my own research on Soloukhin and that subject.

I am deeply grateful to two faculty secretaries who provided considerable assistance: to Missy Price, who typed two of the essays during the early stages of the project, and especially to Eva Owens, who patiently and expertly typed several drafts of the book.

The following persons deserve special thanks: S. Brynn Keith and Markus Pott for their support and encouragement; Vladimir Souloukhin for his kindness and generosity during our meetings in Moscow; Susan Harris, managing editor at Northwestern University Press, for such a rewarding collaboration; and, finally, my mother, Eugenia Caseria, and my husband, Richard Nollan, for their help with the manuscript and for inspiring me in so many ways.

In this book all Russian words and phrases have been transliterated according to two of the systems described in J. Thomas Shaw's *The Transliteration of Modern Russian for English-Language Publications* (New York: The Modern Language Association of America, 1979). For consistency's sake, I have transliterated all pertinent material in the text proper according to System I, for nonspecialists; whereas in the endnotes (my own as well as Souloukhin's) I use System II, the Library of Congress system without the diacritical marks, for scholarly and bibliographical purposes. The only modification I made is the following: in System I, I transliterated the Russian "short *i*" as *i* rather than *y*. I did not transliterate names in the text that are more recognizable to an English-speaking readership by a well-established spelling (e.g., Aksakov, Roerich, Tchaikovsky).

Finally, I chose to *translate* the names of both Russian Orthodox ecclesiastical figures (monks, elders, etc.) and church-related places into the Anglicized forms (e.g., Athanasius, Joseph, the Trinity Monastery of St. Sergius) commonly used by the Russian Orthodox Church Outside of Russia in its English-language publications. This practice should enable specialists and nonspecialists alike to correlate the ecclesiastical names in these essays more accurately with those which appear in writings about Russian Orthodoxy. I hope that the above policy will enlighten, rather than confuse, the reader.

INTRODUCTION

VLADIMIR ALEXEEVICH SOLOUKHIN'S collection of essays, *Vremya sobirat kamni* (A time to gather stones), ranks among the most influential books published in Russian in the post-Stalinist era. The book chronicles various organized attempts of the artistic and cultural elite, joined by people from all walks of life, to save Russia's remaining literary and cultural monuments from utter ruin. In his essays Soloukhin characteristically utilizes literary information as a starting point for the discussion of broader issues, such as the relationship between culture and spirituality, and the causes for the degradation of the environment. He maintains that a country's cultural life provides a barometer or a "hearth" for its people's moral and spiritual sense of well-being, and emphasizes that a country must not forget the sacred task of preserving memories of its past, for without them a people will lose its identity, its place in history, and its link with the future. Embedded in the discourse of these essays is Soloukhin's additional concern about the fate of Russian Orthodoxy, a concern that remained veiled at the time when the essays were written—in the late 1960s and the 1970s, under Brezhnev's ideologically conservative and stagnant regime. Soloukhin couched his thoughts about the state of Russian Orthodoxy in such rhetoric as the need to preserve churches and icons for architectural and artistic (not religious) reasons, and the importance of reevaluating an eighteenth-century poem entitled "Bog" (God) because of its remarkable scientific and cosmological imagery, rather than its clever juxtaposition of the wisdom of the Supreme Being and the nobility of the individual. The essays have become even more important and relevant in light of contemporary Russia's struggle to define her identity and rehistoricize her recent past, particularly the years under Lenin and Stalin. *A Time to Gather Stones* not only represents Soloukhin's urgent appeal to preserve cultural monuments, but also treats indirectly the issue of how to interpret an officially rewritten past, how to unearth the truth about a past that remains buried under layers of propaganda and deception.

Since the 1950s Soloukhin has played a seminal role in contemporary Russian and Soviet letters. His abiding concern about such "burning" issues as the environment, Russian nationalism, and, more recently, the demythologizing of Lenin, has produced an extensive corpus of prose fiction and essays. Born in the Soviet Union in 1924 in the village of Olepino (near Vladimir), Soloukhin became a major prose writer in the 1950s with the publication of his novel *Vladimirskie prosyolki* (Vladimir back roads, 1957), an ostensibly straightforward account of his walking tour through the Vladimir region and his sympathetic observations on peasants and village life. At times, however, his remarks become more subversive than might be apparent at first glance—for example, his seemingly offhand references to churches that are no longer functioning or that lie in ruin.

The novel traces Soloukhin's return as an adult to the pastoral setting of Vladimir, where he had grown up and where his ancestors had lived. In its exhaustive attention to detail and relative lack of action, it resembles the genre of the physiological sketch that was popular in Russian literature during the first half of the nineteenth century. Its tone is informal and lyrical and, like virtually all of Soloukhin's poetry and prose, it is written in the first person. He questions indirectly several fundamental notions of Marxism-Leninism: that a country's past should be erased, that its culture can develop in a vacuum, and that an exploitative attitude toward natural resources will not harm the environment. To be sure, Soloukhin was not the first to express publicly his concern about the fate of Russia and her culture, for many intellectuals both before and after the 1917 Revolution had voiced similar fears: one need only recall the debates between the Slavophiles and the Westernizers in the 1840s and 1850s on the question of Russia's cultural and political future. Soloukhin, however, takes these concerns even further: "He seeks to rouse indignation in his readers, to gather new troops for the struggle against nihilistic indifference toward Russia and her thousand-year history. One thing is certain: Soloukhin's work struck a chord in his readership. 'Several thousand' readers wrote in [in response to *Vladimir Back Roads*—V.N.] asking for more writing on the same theme."[1]

Soloukhin's literary and social activism manifested itself in unequivocal terms in his subsequent collections of essays, which accuse the Soviet government of deliberately seeking to destroy Russia's art, culture, and religion: *Slavyanskaya tetrad* (Slavic notebook, 1965), *Pisma iz Russkogo Muzeya* (Letters from the Russian museum, 1966), *Chyornye doski* (Black boards, 1969), and *A Time to Gather Stones* (1980). In recent years his writing has become even more political in nature, concentrating on the policies that brought misery and death to millions of Soviet citizens during the 1920s and 1930s; this focus on the social effects of Lenin's and Stalin's policies produced the autobiographical *Smekh za levym plechom* (Laughter over the left shoulder, 1988), *Chitaya Lenina* (Reading Lenin, 1989),[2] and *Rasstavanie s idolom* (Breaking with an idol, 1991). Even though Soloukhin continued to publish books of poetry throughout these decades, and even though he considers himself first and foremost a poet, his profound love for Russia and her people, along with his strong moral and religious convictions, expressed themselves forcefully again and again in the essay genre.

To assess the importance of these essays it is necessary to locate them within the matrix of contemporary intellectual/political and literary forces, in particular, those of Russian nationalism and the village prose literary movement. During the period of the late 1950s and early 1960s, intellectuals abandoned orthodox Marxism-Leninism and began to search for alternatives: while some favored a "Western-style liberalism," others considered "variants of Russian nationalism."[3] Moreover, the masses—the tens of millions of believers in the Russian Orthodox faith—felt alienated from the Khrushchev regime by his continued persecution of the church and destruction of religious monuments. However, Soloukhin's novel *Vladimir Back Roads* articulated many of the concerns shared by people of the various social strata in Russia and established "in kernel form many of the themes and motifs which Russian nationalists would evolve over the next two decades[4]": protection of all art and the environment, recognition of the rights of all ethnic groups, criticism of state policies on religious freedom and civic planning, and restoration of the country's historical memory to its people.

Soloukhin's writings reveal that he views nationalism in positive terms, as a natural and essential phenomenon, one that enables a people to cultivate and retain its identity; hence Russia, in reestablishing her traditions, would continue to follow the path of cultural development logical for any society whose people possess a body of shared values and beliefs. If Russia were to lose her link with the past, the loss of memory (represented in part by neglected or forgotten cultural monuments) could prove disastrous to contemporary Russian society by producing a crisis of identity; Soloukhin's essays argue that Russia could very likely deteriorate into a cultural wasteland, particularly since a large part of the people's collective memory—churches, icons, and literary estates—had already been destroyed in Stalin's Great Terror of the 1930s and, later, during World War II.

In his oeuvre Soloukhin's thinking on nationalist issues is remarkably consistent. Whether one peruses his poetry, his short stories, his well-known miniatures *Kameshki na ladoni* (Pebbles in the hand, 1988),[5] or his essays, his stance on Russia's position vis-à-vis other nations and cultures remains clear, to wit: Russia's cultural achievements have been crippled by the ideology of the Soviet State,[6] the Russian people must make a commitment to restore their culture before it is lost altogether, and Russians must maintain an attitude of tolerance and goodwill toward other nationalities.[7] These views locate Soloukhin squarely in the mainstream of Russian nationalist thinking,[8] and moreover within what academician Dmitry Likhachyov considers the best tradition of the Russian intelligentsia of the nineteenth and twentieth centuries, a tradition he calls "spiritual internationalism."[9] Gerald Mikkelson has observed that the essay "A Time to Gather Stones" "is an unabashed apology for a correctly-interpreted Slavophilism, which Soloukhin understands as national pride within a broad international context."[10]

The late 1950s and 1960s witnessed the rise of the literary movement of village prose. The movement's focus on the Russian village as a powerful moral and spiritual force, along with its abiding interest in the prerevolutionary past, converged with the program of the Russian nationalists. Canonical village prose may be a unique phenomenon, but its predecessors exist in Russian literature of the nineteenth century, in the writings of Goncharov, Turgenev, and Tolstoi, among other classical Russian prose writers. Two factors establish a notable place for village prose in the history of Soviet literature: the socialist realist context in which it developed and flourished (during the years of Khrushchev and Brezhnev), and the enormous literary talent of its writers. Village prose evolved partly as a rejection of socialist realism, and partly from its writers' conscious decision to ignore or bypass socialist realism as a theory inadequate for their needs. This literary phenomenon took root primarily because of growing social and ethical concerns among Russian and Soviet readers about socialist realism's whitewashing of the spiritual emptiness of life in centrally planned cities and on collective farms. Kathleen Parthé writes:

> There could not be a greater literary contrast than the one between fast-paced, impatient, future-oriented works of socialist realism, whether about factories or collective farms, and the slow, patient, seemingly passive, past-oriented village prose. . . . To understand canonical socialist realist works one has to see their orientation towards the future, to speed, and to a rejection of the traditional in favor of creating a new world. An appreciation of village prose involves looking at the writers' interest in time, memory, and the past, as well as in qualities such as [spiritual radiance and attachment to one's native soil], and in the image of [the long road], which refers not only to the endless, slow roads of the Russian countryside but also the "long road" back to one's own past and to the past of rural Russia.[11]

Socialist realism by definition opposed the type of writing represented by village prose: it offered psychologically one-dimensional heroes who glorified state policies, and it extolled the virtues of the Party and the collective rather than encouraging the flowering of the individual, whereas village prose characters were average people who pondered moral dilemmas, criticized the state when necessary, and sought fulfillment in traditional values. Moreover, the artistic quality of village prose work, stunningly superior to that of socialist realist writing, attracted the attention both of literary critics and of the general reading public.

How does Soloukhin's work fit into socialist realism's broadly defined canon? Victor Terras describes the following basic tenet: "Socialist realism seeks truth at points where the ideal meets the real, which in practice often means where the present intersects with the future."[12] In light of this notion, the thrust of Soloukhin's work would seem to be all wrong: directed toward the past instead of the future, his essays depict an imperfect present that tragically has lost many of the best features of its past culture, and his authorial/narrative voice correspondingly expresses yearning and melancholy for what has been destroyed. However, in investigating the reasons for this destruction, Soloukhin challenges the socialist realist dictum of the far-sighted wisdom and heroism of central planners and government officials; indeed, he implicates through carefully selected documentation the errors of blundering Soviet bureaucrats who neither understood not cared about preserving monuments of the past that have shaped Russia's cultural identity. According to socialist realism, Soloukhin's work is ideologically incorrect in its emphasis on the so-called setbacks, rather than the achievements of the Soviet state. Moreover, Soloukhin treats these setbacks as major national tragedies rather than as actions that pale in importance when perceived in light of the relentless and all-engulfing drive toward the future.

In the essay "Aksakov's Territory," perhaps in an attempt to meet one of the requirements of socialist realism (that of art's service to the state), Soloukhin masterfully interweaves the significance of preserving monuments of the past with the patriotic feeling (*chuvstvo rodiny*) that future generations of Soviet children will experience upon either reading Aksakov's work or visiting his estate. The salubrious effects of the Russian countryside surrounding the estate, Soloukhin argues, will intensify the children's identification with the Motherland. Soloukhin defines this patriotic feeling as follows: "But this 'feeling for one's homeland' is very complex, for it includes a feeling for the history of one's own country, a feeling for the future, an evaluation of the present, and in addition—but not last in priority—a feeling for nature in one's homeland." Here Soloukhin's deep conviction about the sources of patriotism coincides with the social role of the writer, as determined by the proponents of socialist realism, and also with one of the features of Russian nationalism—love for one's country and its traditions.

As a creator of both poetry and prose, Soloukhin differs from most other village prose writers in the fact that he incorporates the present into his work as a worthy subject, rather than concentrating exclusively or primarily on apocalyptic events of the past, such as forced collectivization under Stalin and various Soviet environmental disasters, and the devastating effect of those events on the Russian village. His desire to examine Russia's struggle to preserve the essential features of her own culture leads naturally to matters of art, such as the distinction between genuine and mediocre poetry, and the relationship between beauty (or ugliness) and the condition of the soul. In his reflections on such issues as the origins and meaning of art, its relationship to religion, and its effects upon society, Soloukhin moves increasingly further from socialist realism. Because he addresses the needs of the inner, spiritual life that thirsts for beauty, truth, and religious awareness, he correspondingly moves away from serving the stated needs of a particular society. In becoming more universal, his artistic vision (concerning the relationship of the artist to society at large) becomes less ideologically tendentious.

Soloukhin clearly considers the Russian Orthodox faith and enduring works of art and literature to be essential components of Russian culture and to represent the hope for Russia's future. Thus the preservation and restoration of cultural monuments becomes a meaning-laden act, closely related to respect for the traditional rural environment and to Soloukhin's own religiosity. However, neither the religious beliefs that can be inferred from his essays nor his crusade for the art of the past has brought about his estrangement from socialist realism; instead, what could be called his "excessive preoccupation with the loss of village life" stands as antithetical to socialist realist doctrine. In Soloukhin's work progress is often viewed with suspicion, because it may irrevocably erase the past, to the detriment not only of the Russian people, but of all humankind.

Although Soloukhin writes in a straightforward manner that eschews unnecessary ornamentation, his style is both informal and sophisticated. Many Soviet and Western literary critics consider him one of the most talented stylists in contemporary Russian and Soviet belles-lettres; David Lowe has noted, "Soloukhin writes a poet's prose, and his leisurely pace and lush language have earned him the high regard of Russian readers, many of whom regard him as one of the most eloquent and graceful writers of the age."[13] Indeed, the essays of A Time to Gather Stones have thus far defied generic categorization, for in them "reportage is interwoven with penetrating lyricism, the documentary with generalization."[14] Splendid examples of Soloukhin's bantering yet elegant style, the essays are witty, humorous, inspiring, and ultimately tragic. They contain much sophisticated humor as well as the more prosaic, hilarious remarks Soloukhin aims at himself and his self-proclaimed scholarly inadequacies. His essays in particular successfully fuse his nostalgia for the past with a genuine interest in the present and future, thereby satisfying a key criterion attached to the image of the model Soviet writer, according to Kuznetsov, namely, that his or her works manifest "activism and the social responsibility of the author's position, a tendency toward direct involvement in life."[15]

Soloukhin's essays share certain distinguishing features, which fact in part accounts for the difficulty of establishing their generic boundaries. Their elegance and informality is accompanied by a density resulting from the multifarious modes of expression Soloukhin interweaves into the text: the insertion of entire documents (letters, resolutions, remembered speeches) and quotations (from newspaper and encyclopedia articles, and from Russian literature); the alternation of dialogue and first-person narration, with occasional rhetorical questions to the reader; and the enumeration of names and places for their rhetorical, emphatic, and incantatory effects. Even though Soloukhin's prose is not strictly poetic, occasionally the poet emerges, such as in his comparison, in the essay "Greater Shakhmatovo," of people congregating around the stone at the Blok festival to bees gathering around their queen. But regardless of the approach Soloukhin takes in his writing, his shifting voice (as author, narrator, and/or occasional character) predominates and remains in control. Because Soloukhin's subject matter is both autobiographical and civic-minded, the reader never loses sight of the narrative voice that alternately chides, chuckles, reproaches, laments, and mourns.

The collection *A Time to Gather Stones* consists of four essays published during the 1970s and a paper presented as part of the Nobel Symposium in Göteborg, Sweden, in 1978.[16] The latter, "Tsivilizatsiya i peizazh" (Civilization and landscape), manifests a thematic and philosophical confluence with the other four essays in its link between beauty and morality, and the effects of uninspired art and architecture on the human spirit. The impact of the five essays on the reader is both individual and cumulative; each part is as strong as the whole. Their powerful impact derives in large part from Soloukhin's rhetorical strategy—as in *Vladimir Back Roads*, his writing is more subversive than may be readily apparent. In the second essay, "Aksakov's Territory," the Aksakovs' house serves as one of the collection's central metaphors: it was torn down with great difficulty, log by log, in the same way that the Bolsheviks attempted to eradicate Russian culture, monument by monument. Hence the title *A Time to Gather Stones*, in addition to its biblical association (Eccles. 3:1–5), must be understood in its cultural context as well: just as some primitive societies employed (and may still employ today) a system of counting goods or livestock by associating one stone with one object, so Russian society, after the disastrous experiment of communism, must begin the process of tallying its cultural monuments. Soloukhin's essays resonate with the meaningful collective ritual of assessing, in numerical and other, more sophisticated terms, the condition of a country's historical memory.

The publication history of the essays is as follows: they first appeared as a collection under the title *A Time to Gather Stones*;[17] the collection was subsequently incorporated into two, more comprehensive series of essays by Soloukhin, which were published under different titles: *Letters from the Russian Museum* and *A Time to Gather Stones*.[18] Soloukhin has noted that the essays of *A Time to Gather Stones* were republished each time without significant revisions, except for minor editorial changes; he also affirmed that they were based on facts and actual experiences but reworked subsequently into literary (essay) form.[19] The four essays are offered here in their original chronological order, followed by the paper "Civilization and Landscape."[20]

Although the four essays of the collection differ in the specific elements Soloukhin introduces (official documents, lines of poetry, quotations from ecclesiastical texts) in order to construct his argument, they have in common the following basic structure: a lively introduction, often containing a personal anecdote, followed by the historical and biographical background pertinent to the cultural monument(s) in question, and culminating in a visit by Soloukhin (typically accompanied by friends—art historians, writers, or other members of the intelligentsia) to the site of the monument(s): Derzhavin's estate, Zvanka ("A Visit to Zvanka"), Aksakov's estate in Buguruslan ("Aksakov's Territory"), Blok's family estate, Shakhmatovo ("Greater Shakhmatovo"), and Optina Monastery and Shamordino Hermitage, associated with such writers as Gogol, Dostoevsky, and Tolstoi ("A Time to Gather Stones").[21] Soloukhin's description of the ruined monument(s) builds to a heightened state of thoughtfulness and cultural awareness, and proceeds to his moving, melancholic closing remarks. The paper "Civilization and Landscape" successfully completes the collection in its broadening of the focus of the preceding essays to include a discussion of the aesthetic categories of beauty and ugliness in their relationship with the external appearance of Earth—through both natural and human-made wonders—and the effects of landscape, art, and architecture on the human spirit.

A Time to Gather Stones may be thought to constitute, in Mikkelson's words: "a prescription for the exalted status of the 'true poet' (*istinny poet*) who must possess an acute sense of the word (*oshchushchenie slova*), the courage of his convictions (*smelost*), close ties to his native soil and culture, a sense of the continuity between past, present, and future, a willingness to sacrifice his own earthly welfare for the defense of 'imperishable values' (*neprekhodyashchie tsennosti*), and an ultimate allegiance only to his own 'soul.'"[22]

In his collection of essays on the nature of art, *S liricheskikh pozitsii* (From a lyrical point of view),[23] Soloukhin states the following paramount goal of art: to make the heart a little kinder ("Only art can do this, and moreover, this is its main, everlasting goal"). If science is the collective memory of reason, of the intellect, then art is the memory of feelings, of emotions. Art becomes the collective memory of humankind, and the artist should remember that future generations will perceive his or her art as embodying the spirit of the times in which he or she lives. Because of this moral responsibility toward the future, Soloukhin charges each artist to strive for an art that is honest and ennobling to the human spirit.[24] In the case of *A Time to Gather Stones*, perhaps the most important quality of the essays, in addition to their profundity of thought and literariness, is Soloukhin's sincerity—what Mikkelson has called his *"podvizh-nichestvo,"*[25] his selfless devotion to an honorable cause. This may be Soloukhin's—and his essays'—greatest gift to all of us.

A VISIT TO ZVANKA

"ZVANKA? WHAT'S ZVANKA? Probably some sort of run-down monastery, isn't it? Or maybe a progressive state farm?"

We were traveling down the Volkhov on the hydrofoil *Raketa* (Rocket) and at any moment expected to be dazzled by a blue sea, not a southern or a Crimean or an Adriatic, but rather a northern blue. Lake Ilmen. Sadko. The reddish-colored Viking longboats of Roerich[1] on the blue. Blinding-white fleecy clouds. And in addition, level deep-green shores. Everyone knows that a southern sea is always accompanied by cliffs, pebbly sand, and yellow steppe. Dry wormwood, willow herb, tumble-weed. Dry barrows with eagles perched on them. Here the verdure is as lush as on a floodplain. Here there are daisies, buttercups, pink knotweed. And if you see a rock, it will be a rounded boulder that brings to mind a heroic epic and that presages a vision, not of a Pecheneg or a Tatar, but of a Viking in armor who has taken off his helmet, revealing blond curls on iron shoulders. Rus.[2]

We, a group of Moscow intellectuals, had gathered at that time in Novgorod for a conference marking the millennium of culture in this city. Actors, philologists, architects, writers, archaeologists, artists, musicians. Enthusiasts. There were speeches, reports, and resolutions. And there was a cruise along the Volkhov on the *Raketa* headed for Lake Ilmen, blue with its level, deep-green, daisy-covered shores. With its rounded boulders and the gray cottages of villages, in which all of the best riddles, tales, songs, sayings, and heroic epics had been born and preserved over the centuries.

We were expecting to be dazzled at any moment by a view of the lake, when suddenly the little and somehow familiar (we *were* Moscow intellectuals!) word "Zvanka" came up.

It didn't happen accidentally. It was I who sent it out "to the masses" on board the *Raketa*, or, more precisely, not I, but my friend Volodya Desyatnikov, at my request and with my encouragement.

Desyatnikov is a man of action, a man who gets down to business. And this is the right approach in our twentieth century, for we talk a great deal but take a long time to prepare for action. Sometimes it all ends this way, in talk and meetings. We put it all off until the next time, until next year, but in reality, until forever.

Take me, for instance. I always dreamed of spending some time at Zvanka, but now, upon finding myself in Novgorod, I decided: the next time I come here I'll definitely plan to visit Zvanka. In all honesty, I did make certain inquiries: I stopped by at the local newspaper office and asked where Zvanka was and how to get there. I was told that I would need an all-terrain vehicle, because to travel the entire way along the banks of the Volkhov would amount to a distance of seventy kilometers. I wanted to be sarcastic and ask which all-terrain vehicles the landlord of Zvanka had used to get there in the eighteenth century, but I held my tongue. It turned out that the newspaper office chauffeur had become ill, and my plan began to wither away at the roots. I didn't try to insist or look for other ways to get there, but said to myself: next time.

Finding myself aboard the *Raketa*, and without ulterior motive or malicious intent, I confided to Volodya Desyatnikov:

"Of course, it's not a bad thing to take a ride along the Volkhov even without a purpose. But why not make use of the *Raketa* and visit Zvanka? In order to do this, we must all have the desire and we must all be in agreement. Could it be that the others would not also find it interesting to see Zvanka? The 'masses' can be persuaded. One must mention (introduce) the idea, create popular opinion, and then make a suggestion. The *Raketa* will have to turn around and travel in the opposite direction. But isn't it all the same to the *Raketa*, no matter which way she goes?"

Volodya instantly supported my idea, and we formed something akin to a conspiracy. We heard people talking on the *Raketa* about Lake Ilmen, about the Novgorod Kremlin, about the restorers Grekov, about everything imaginable, only not about Zvanka. But the experiment had already begun: Volodya had already moved away from me and had merged with "the masses," and after no more than five minutes, as if of its own volition, coming out of nowhere, the little word "Zvanka" appeared and began to resound.

"Zvanka? What's Zvanka? Probably some sort of run-down monastery, isn't it? Or maybe a progressive state farm?"

"Zvanka? What do you mean?! Zvanka is the place where Derzhavin lived, his estate. Lermontov owned Tarkhany; Pushkin, Mikhailovskoe; Tolstoi,

Yasnaya Polyana; Turgenev, Spasskoe-Lutovinovo. . . . And Derzhavin, Zvanka."

"Oh yes, of course! I remember. He wrote a poem containing a description of Zvanka. What is it called, does anyone remember?"

"'To Evgeny. Life at Zvanka.' One of the remarkable lyrical works of eighteenth-century Russian poetry, idyllically depicting life on a country estate in all its detail. The poem describes the daily routine of Derzhavin at Zvanka, on the banks of the Volkhov. Household affairs, an illiterate estate elder, hunting, walks, work, entertainment. . . . 'Life at Zvanka' is dedicated to Metropolitan Evgeny Bolkhovitinov, the compiler of a dictionary of Russian secular writers, an archaeologist and historian of Russian literature. He lived in Khutynsk Monastery about sixty versts[3] from Zvanka and was a friend of the poet during the last years of his life."

"Yes, of course. I remember. I've even seen a gravure depicting Zvanka somewhere. I think there was a high mountain. . . ."

"Not a mountain, but the steep bank of a river."

"What river?"

"The Volkhov, of course, the river we're traveling on right now!"

" . . . And it seems there was a wide stairway leading from the water to a two-story house. A park and some other buildings were near the house."

"But that's somewhere right near here!" the triumphant conclusion dawned on one of the passengers.

"Yes, it's here on the Volkhov. I wonder if it's far away."

"An hour and a half there, an hour and a half back." Volodya Desyatnikov turned up right at this moment. "Along the picturesque banks of the Volkhov."

He reduced the distance a little, but at such a critical moment one had to underestimate. If he had said "seventy kilometers," the passengers would have started thinking and would have wavered: maybe we should postpone the excursion until the next time.

Everything Volodya Desyatnikov had said couldn't have produced better results. From out of nowhere came the suggestion to cancel the ride on Lake Ilmen (on the whole, a purposeless ride over a watery expanse) and go to Zvanka. The *Raketa* swung round and from the very gates, one might say, of the lake moved along the shimmering Volkhov, the ancient Volkhov, where once the sig-fish and the tsarevna Volkhová had lived, and where now can be found, for example, the Volkhovstroi[4] (even farther than the supposed Zvanka) and vessels called *Raketa*, while along both banks one sees nothing but ruins, nothing but dilapidation. Here one sees the ruins of Khutynsk Monastery, there the ruins of the Arakcheev Barracks, and farther on, the remains of an unfamiliar country estate whose name remained unknown to those of us on the *Raketa*. Heavy fighting took place here on the Volkhov.[5] But the ruins make a powerful impression, even in their present state. The maxim is well known: magnificent architecture is magnificent even in ruin.

Khutynsk Monastery (where Derzhavin's friend Evgeny Bolkhovitinov had labored) appeared off the starboard side when we had barely set out from Novgorod. It stood out from the midst of the luxuriant wild vegetation because of its crumbling brick walls and the jagged silhouette of its church with a cupola that had fallen on its side. All that remained of the cupola was its iron framework, as is often the case with church cupolas. Their latticework characteristically appears against the background of the sky.

Meanwhile, there remained the question of the one and a half hours, actually two hours (and two hours *are* two hours); since it wasn't permitted to stand on the deck of the *Raketa*, one could only sit in the glassed-in lounge. This is why everyone soon got tired of contemplating the riverbanks and joined in the general conversation, which naturally kept returning to the topic of Derzhavin, since we were going to visit him, in the same way that we might visit Lermontov at Tarkhany, Pushkin at Mikhailovskoe, Esenin at Konstantinovo, Blok at Shakhmatovo, or Tolstoi at Yasnaya Polyana. . . .

"Here's an example," someone was saying, "of how lazy and lacking in curiosity we are. We know hardly anything about Derzhavin, even though the bare facts of his biography already reveal a fascinating tale of how an ordinary soldier of the eighteenth century rose to the highest ranks of the government, and became a governor, the private secretary of Catherine the Great,[6] and a senator."

"It's true that the sovereign (Her Majesty) was not entirely satisfied with her confidant's behavior. She would complain, 'This gentleman is shouting at me!'"

"The uniqueness of the case lies in the fact that this man rose to the highest government offices, not by official, or political, or diplomatic and military capabilities, but because of his poetic gift."

"He was a courtier and a flatterer!" a seemingly strange voice cut into our friendly conversation.

"That's nonsense."

"Why is it nonsense? Wasn't it Derzhavin who penned the ode 'To Felitsa,' in which he glorified Catherine?"

"It's nonsense because Derzhavin was a genuine poet. He didn't know how to play the hypocrite. On the contrary, he was bold and too straightforward. While frequenting the company of princes and nobles, he would suddenly declare calmly to their faces that neither a coat of arms nor the shadows of ancestors make a nobleman a member of the nobility. How do the lines go?

> I am a prince if my spirit shines;
> A master if I master my passions;
> A boyar if I bleed for everyone, . . .[7]

These lines are pleasant for us to read today, but just try and imagine writing them in those days. I think that at that time it was outright sedition."

"Having praised the empress, however, he could allow himself some political frivolity concerning her noblemen. 'Felitsa' was like a safeguard for Derzhavin."

"'Felitsa' was one of Derzhavin's early poems. It's common knowledge that he didn't even want to promulgate it. For more than a year it lay in his study, until a certain Kozodovlev, who was employed at the Academy of Sciences and who lived in the same apartment house as Derzhavin, accidentally came across it, asked to borrow it for a day, and then spread the word."

"Everything you say doesn't mean a thing. The fact is, that when writing an ode Derzhavin expressed his bona fide thoughts and emotions in it. He did not play the hypocrite."

"How do we know this?"

"Subsequently he was told many times that Catherine was expecting another ode, but the poet was unable to squeeze out a single word. This confirms the fact that a genuine poet cannot write something that would compromise his soul. Memoirs have survived of the protracted torments of the poet, who labored in vain to produce even a single sound from his lyre. 'No matter how many times he would set about the task, sitting for weeks at a time locked in his study, he was not able to create a work that would satisfy him: It all appeared lifeless, forced, and ordinary, like the work of other commercial poetasters who produce only words, instead of thoughts and feelings.'"

Stop. Stop. Let's linger for a while on this point. After all, this question concerns not only Derzhavin's behavior, but also the psychology of the creative process in general. I know a talented poet who published little and consequently was forced to work as a reporter for a newspaper. People made fun of him. It would take several days for him to squeeze out the simple little articles that others could dash off in forty minutes. And you know what? His articles turned out worse than those of the other, less talented people. Having understood this, they no longer assigned him articles, but rather poems commemorating holidays or special events. In honor of International Women's Day, for example. There are many special days that are celebrated. But now from his pen there issued pathetic, lackluster, unoriginal words. It was as revolting to read his poems as it would be to eat food that somebody else had already chewed. It was terrible. But at the same time the man had talent, and when he created his own poetry everything was instantly transformed. It resounded like bronze, and glittered like pure gold.

So, if Derzhavin had been a mere poetaster, he would have composed dozens of these useful odes in praise of Catherine. But he couldn't do this because he was a genuine poet; in a quatrain Derzhavin himself expressed perfectly the predicament of poets when others force them to write poetry:

> They caught a little songbird
> And clenched it in their hands.
> It could not sing, but only squeaked;
> And they said: "Sing, bird, sing."[8]

"But, in general, it's now practically impossible to read Derzhavin's verse, with its ponderous phraseology and his eighteenth-century vocabulary."

(I must apologize, perhaps, for the fact that our conversations are rendered here in a so-called revised form. They were, of course, livelier, more fragmented, and less logical. The quotations, especially those in prose, couldn't have been this precise and complete, despite the fact that only well-known specialists were on this trip. In recalling which aspects of Derzhavin's poetry our conversations touched upon, I subsequently referred to the texts, because at this point the essence of the matter was more important than a reporter's accuracy.

Moreover, in remembering our trip and conversations, I unwittingly added some thoughts, while some directions that the conversation took seemed incomplete and were expanded further when I was alone, sitting over a sheet of paper.)

"Who told you that?" immediately responded Derzhavin's defenders. "In places his style really is unwieldy. But as soon as you penetrate the purple-velvety, golden blanket of his verse, you find yourself in a world of striking visual, very concrete and earthly imagery! In addition, in places Derzhavin's style achieves the lightness and elegance of Pushkin, who must be considered the direct heir and successor of Gavriil Romanovich. In which way, for instance, does the following stanza not evoke Pushkin?"

> And if I am beloved
> By my dear and sweet Plenira
> And in Fate's changing tides
> I can boast of faithful friends,
> If I live in peace with neighbors,
> And can sing and play the lyre
> Then who is happier than I?[9]

It's completely puzzling that the major critical topic "Derzhavin and Pushkin" or, if you will, "Pushkin and Derzhavin" has not been studied at all. In leafing through the prodigious volumes of Pushkiniana, we find the following sections: "Pushkin and Ossian," "Pushkin and Parny," "Pushkin and Chenier," and

"Pushkin and Byron." But not a single line has been written on the theme "Pushkin and Derzhavin," except for an account of the bare fact that Derzhavin heard the young graduate of the Lyceum read and the line by Pushkin himself: "The old man Derzhavin noticed us and, so close to death, gave us his blessing."

But meanwhile, Pushkin knew the work of his predecessor thoroughly, though not by heart, and he held it in high esteem. In a letter to A. A. Bestuzhev, while answering Bestuzhev's questions "in paragraph form," Pushkin writes: "Why do we have no geniuses and few people with talent? First of all, we do have Derzhavin and Krylov. Second, where on earth can one find many talented people?"

But one can judge the kinship between these two poets best of all by the similarities in their intonation, subject matter, and the structure of their imagery. For example, take these two stanzas:

> The music plays; choirs sing
> Near your gourmet tables;
> Mountains of sweets and pineapples
> And a multitude of fruits
> Entice your senses and feed you;
> Young maidens serve you,
> And offer a stream of wines:
> The aliatiko and champagne,
> Russian beer and English,
> Moselle and seltzer water.
> —Derzhavin[10]

> He saw: corks fly to the ceiling
> Streams of wine spurt like comets,
> Before him was roast beef very rare,
> And truffles, youth's luxury,
> The best of French cuisine,
> And fresh Strasbourg pie
> Between ripe Limburger cheese
> And golden pineapple.
> —Pushkin[11]

Let's take one more example:

> A pot of good, hot cabbage soup,
> Smoked ham over the fire;

> Surrounded by my family,
> To whom I am the master,
> That's why my dinner tastes so good!
> —Derzhavin[12]

> My ideal is now—a wife,
> My wish—for calmness,
> Some cabbage soup, myself the master.
> —Pushkin[13]

I'm not a literary critic: I don't intend nor do I know how to analyze texts. But if such a science exists—namely textology (is there such a science?)—then it wouldn't hurt for scholars specializing in it to undertake a comparative analysis of the texts of two of our greatest poets, one of whom is rightly acknowledged and considered to be great, while the other has been unjustly relegated to secondary-school readers, and even then, to standard passages learned by heart, such as "The divine tsarevna of the Kirgizkaisatskaya horde, whose wisdom is incomparable" or "Where the table was full, there stands the coffin."[14]

It's true that in his article "G. R. Derzhavin. Life and Works" A. Ya. Kucherov plainly states, "Derzhavin ranks among the greatest Russian poets," but this statement is somehow constructed so that "ranks among the greatest" indicates that there are many great poets, and consequently the word "greatest" in the given context sounds weaker than if only the word "great" had been used.

To be sure, Derzhavin's poetry is somewhat ponderous. His contemporaries knew this before us. More likely, not his contemporaries, but his first and immediate successors. Bestuzhev, Pushkin, Gogol, and Belinsky concurred that, on the one hand, Derzhavin was a genius but, on the other—virtually illiterate.

"Formidable virtuosity suddenly turns into untidiness. . . . If they had properly educated such a man, there would not have been a poet superior to Derzhavin" (Gogol).

"Here one finds a real lump of unrefined ore containing bright flashes of pure gold" (Belinsky).

"His style is as elusive as lightning. But his rapture in its flight often overlooked grammatical rules, while mistakes burst through along with beauty" (Bestuzhev).

The poems of Derzhavin, "despite the irregularities of their style, are full of flashes of genius. His daring, his supreme daring" (Pushkin).

It was Pushkin who made the following startling assessment of Derzhavin's verse: "When reading his work, one feels as if one were reading a bad translation of a marvelous original."

In Pushkin's time the concept of an interlinear translation did not exist. Writers knew foreign languages. They read Byron and Schiller, Voltaire and Shakespeare in the originals and translated directly from them. This was the case even with Homer. And if Lermontov produced a poem subtitled "From Goethe," then one could confidently assume that no third person had done an interlinear translation for him.

For us the notion of interlinear translation is as common and ordinary as that of a fountain pen or tape recorder. This is why, Pushkin's striking formulation notwithstanding, Derzhavin's poetry has always seemed a little like an interlinear translation to me, one that elicits the desire to recast (translate?) it into a lighter and more contemporary (Pushkinian?) language. But these are only two sides of the same coin.

What is amazing is the following: Can it be that all those who evaluated Derzhavin's poetic style and who reproached him for his style did not remember that he had lived before them, did not know them, and had not read Pushkin, or Gogol, or Belinsky?

What kind of marvelous original did Pushkin envision in Derzhavin's ponderous stanzas? Probably an original written in the Russian poetic language of Pushkin's own time. But, after all, Pushkin in fact created and consolidated this language. Hence, without any clever wordplay, we can state that Derzhavin's verse is a "bad, free translation" from the poetic style of Pushkin. Or, in reverse (if we consider Derzhavin's poetry as an interlinear translation, i.e., as raw material for a translation), Pushkin is a "wonderful translation" from the poetic style of Derzhavin.

These days it is easy for us, who gracefully fly at the speed of sound, to pass judgment on the first unwieldy airplanes. We want so much to round off their rectangular wings, streamline their fuselage, make their landing gear retractable, and install a different engine.

But if we take this technical analogy, then it's still impossible to equate Derzhavin's poetry with the first airplanes. The miracle lies in the fact that the apparently clumsy airplane of Derzhavin's verse suddenly soars upward to heights and distances that couldn't be attained subsequently by Russian poets, neither in Pushkin's time, nor in Nekrasov's, nor in Blok's, nor—especially—in most recent times.

Antokolsky (during a speech at the Writers' Club, I don't recall on what occasion) said: "Here is a poem that could appear at the beginning of an anthology of Russian verse." He was referring to Derzhavin's poem "Bog."[15] Of course, there was Lomonosov with his discourse in verse on the usefulness of glass, and Trediakovsky with his *Telemakhida*.[16] If one made choices based on accuracy and scholarly objectivity, then it would be impossible to do without Kheraskov, or Kantemir, or Kapnist. But Antokolsky probably had in mind a limited anthology, an anthology of exemplars that could function without his-

torical order or the principle of representation, an anthology whose sole organizing principle would be that of significant and genuine poetry.

To be sure, even in this case I would preface Derzhavin's poem with a stanza by Trediakovsky. This man, whom Peter the Great[17] identified as a worka-
holic when Peter saw him as a child; this man, who all his life manipulated the
awkward stones and millstones of his *Telemakhida* (Catherine the Great required
her officers to memorize *Telemakhida* as penance, instead of sending them to the
guardhouse), out of which only a single line is well known now, because
Radishchev used it for the epigraph to his *Journey* ("The monster is grim, gigantic, savage, hundred-mouthed, and bellowing");[18] this same man suddenly
invented a stanza full of power, energy, and charm:

> Hark, oh sky, I proclaim,
> The earth should hear my words:
> My words will flow like the rain
> And, like dew to a flower,
> My deeds descend far and wide.[19]

For a long time I have dreamed of compiling for myself "my own" anthology of
Russian poetry, that is, those poems which I personally like. Perhaps in the
process I would eliminate the entire corpus of poets who have earned a place
in the history of our literature; perhaps, conversely, I would select several
poems by poets completely outside of that commonly known chain in which
our poetic names are extended and threaded, and which, unfortunately, lacks
many, many links.

I, too, consider Batyushkov, Ryleev, and Delvig to be poets who are very
important for the development of our native poetry, since Pushkin did not
mature in a desert and Pushkin alone does not constitute the entire Pushkin era
anyway. I understand this, but even so, perhaps I would not include a single
poem by Batyushkov in "my" anthology.

I also realize that Polonsky, God knows, isn't a great poet, that he is not a
Nekrasov or a Blok. However, I know and love several poems by Polonsky,
and when I recite them from memory to my friends, they are all astonished and
ask who wrote such a wonderful poem. And I answer that it was Polonsky.

In the case of compiling a subjective anthology of Russian poetry, I think I
would open it with the stanza by Trediakovsky cited above—specifically, with
the stanza and not the whole poem. As for Antokolsky, he considered it possible to begin an anthology with Derzhavin's poem "God."

Here is additional evidence of the fact that Derzhavin was a legitimate
poet. He conceived his mighty poem, his ode, in 1780. Not only conceived,
but began work on it. However, for four years the work did not progress.

"Being preoccupied with his official duties and various social obligations, he could not finish it, no matter how hard he tried." That's how it was; and yet, that's not how it was. Neither official duties nor social obligations prevented Derzhavin from writing several remarkable poems during this four-year period, including the well-known, illustrious "Felitsa."

Poets themselves often compare the gestation of a poem in their consciousness with the maturation of fruit. The analogy works precisely because in both cases the process is completely unconscious. In other words, the poet is preoccupied with official duties and social obligations, and meanwhile the poem continues to ripen. Then comes the time to give birth. It is impossible to stop the process.

In 1784, while traveling through Narva, Derzhavin left his carriage and his servants at an inn and then disappeared for a long time. It turned out that he had rented a small room from an old woman, had sequestered himself there, and during those few days had written down what had been ripening for four years. Indeed, as we would say today, "the time had come."

The fact in and of itself is already noteworthy. What other poet would make this kind of stop, in the middle of a journey, to sequester himself at an old woman's place and write? After all, social obligations probably awaited him, too, if not official duties. But all this has fallen away and been forgotten. Only the genuine creative work, the product, remains. The ode has been translated many times into all the European languages, and consequently (according to specialists) is even better known in other countries than here at home; this can be explained by the fact that, after all, it would be difficult to popularize a poem entitled "God" among Soviet primary-school students. What does it matter that Derzhavin himself had in mind "endless space, endless life in the motion of substance and the eternal flow of time"? For those of us who have grasped the fundamental tenets of materialism, such as "matter is primary, consciousness secondary," it is easy to make determinations about the endlessness of space and the eternity of time. This is not the eighteenth century, the time when Derzhavin wrote. But still, it makes one think sometimes. We acknowledge the secondariness of consciousness. So be it. And by doing this we already separate consciousness from matter and elevate it to its own level. I mention this because, when entering the labyrinth of philosophy, it helps to establish the terminology: we say "matter," Derzhavin said "substance." We say "consciousness," Derzhavin—"spirit."

The very endlessness and eternity of the world, universal laws to which all things are subject, beginning with the solar system and ending with the atom (and the solar system itself is no larger than an atom on the scale of the universe, while the universe itself is no larger than an atom on the scale of . . .). In short, the very endlessness and eternity of the world and the unity of laws to which all things are subject, which penetrate everything and set it in motion, had to be called something. In Derzhavin's time people said—"God." The poet, then, simply begins his poem this way:

> Oh, Thou, of endless space,
> Who livest in the motion of substance,
> In the flow of eternal time, . . .

That is, from the little room rented at the old woman's in Narva, he glimpsed such an abyss that it seemed as if time had existed before the dawn of time, eternity before eternity.

> Oh spirit, who existest everywhere, indivisible,
> Who art beyond place or reason, . . .

It is evident that the poem is not religious or ecclesiastical at all. It contains neither the cloud with Sabaoth sitting on it, nor Hell, nor Paradise, nor the prophet Elijah in his chariot.

> Who art beyond place or reason,
> Of whom we cannot conceive,
> Whose spirit fills all things,
> Encompasses, builds, and preserves,
> The one, whom we call—God.

Yes, the word is uttered. But, I think, with the utmost clarity. It would have been possible to settle on another term for what the poem describes. The essence would not be altered. Time is endless, space boundless, motion perpetual, the central laws of motion unfathomable— i.e., we cannot penetrate and comprehend the causes and effects of existence; the central point, around which in the final analysis everything moves, is not known to us. One must, after all, attach some word or other to this point!

But, in all honesty, I'm interested professionally not so much in the philosophical aspect of Derzhavin's ode (actually—narrative poem) as in its composition, its structure. I remember, and this was in my youth long ago, that when I read the poem for the first time I was instantly struck, not by the vastness of its conception, but specifically by its structure. I was overcome with a kind of rapture when I noticed the concentric circles, at first growing ever smaller until they diminished to a point of infinitely small size, and then suddenly from this point new concentric circles beginning to grow. Only a great master could incorporate such a precise form into his artistic work; this is why

I cite "God" as an example of Derzhavin's purely literary craftmanship. Let's attribute everything Derzhavin says about God, about substance and spirit, to his eighteenth-century lack of information, looking instead at how he constructs, crafts, creates, and organizes. Of course, in order to understand Derzhavin better, we shall have to assume his views for a few minutes. But afterward it will be easy to reject them and return once again to our informed, enlightened, and sober state of mind. Derzhavin placed the Individual and God face to face and began to compare them apropos of time, space, greatness, and even their very size. The contrasted "I" and "Thou" stand facing each other.

The first stanzas are intended to show the insignificance of the first as compared with the sublimity of the second. In these stanzas Derzhavin is preoccupied with self-abasement and strives for the complete deflation of himself, that is, of humankind in general. Concentric circles, at first of cosmic dimensions, keep contracting until what remains in the center is something that is impossible to discern but that nevertheless can be called "I." Here is how this takes place in the poem. It begins subtly:

> To measure the deep ocean,
> Count grains of sand, the rays of planets
> Is possible for the human mind,
> But Thou art beyond number or measure!

The poet chooses ordinary and comprehensible things. It is true that the depth of the ocean remained a mystery in the eighteenth century, but yet to measure it was a realistic and conceivable notion. To count the stars? Possible, although more difficult. To count the grains of sand on our planet would probably be even more difficult, although there are and must be a finite quantity and limited number of grains of sand; but in the line "Thou art beyond number or measure" lies the first, though tentative, determination of the scope of "Thou."

If we take a closer look, we shall also see that the ensuing poem is an enlargement of the first stanza and serves as a commentary upon it. The poem grows out of this first stanza, like a luxuriant tree from the smallest seed. It seems as if, in this stanza, everything has already been stated, everything has been presented, and the poet only needs to expand, elaborate, and convey by means of literary devices. For example, the inconceivable is mentioned ("of whom we cannot conceive") and later expanded:

> Enlightened spirits
> Who from Thy light are born

13

Cannot explain Thy destiny;
Only thoughts dare to reach Thee,
And disappear in Thy greatness,
Like a passing moment in eternity.

Or it is mentioned that God is "beyond place or reason," that his spirit "fills all things," and here is the expanded commentary:

The chaos before the dawn of time
Thou called from eternity's abyss,
Eternity, born before the ages,
Thou created in Thyself.
What Thou art exists in Thee,
Thy radiance shinest in Thee
Thou art the light, whence light was born.

When you try to imagine everything Derzhavin wanted to say, it makes your head spin all by itself. Eternity that existed before eternity, time before the dawn of time, light before the appearance of light, radiance that shines from out of itself. . . . But let us return to the graphic, purely visual, concrete scope of "Thou":

Like the flying and rushing of sparks,
So suns are born of Thee; . . .

What sparks could Derzhavin have seen in his day? He probably didn't have in mind the output of fused cast iron from a blast furnace. Sparks from a bonfire, a conflagration? Sparks from under a blacksmith's hammer, when the red-hot iron is being forged on the anvil? Let's imagine one scene and then the other. The fire blazes, while a multitude of sparks flies up into the dark of night, gradually becoming more diffuse, more rarefied, and finally disappearing. They fly into the air without any particular effort, easily, lifted up by the fiery air. From under the hammer the sparks scatter like arrows, in bunches, in spurts. Let's imagine that in a similar fashion it is no longer sparks that fly in all directions, but suns; now we can understand approximately what the poet wanted to convey to us. But even this image seemed incomplete to him. Would the sparks fly a great distance? It becomes necessary to consider something of greater scope:

As on a frosty but clear day in winter
When specks of hoarfrost glitter,

14

> Whirl, surge, and sparkle,
> So under Thy gaze are stars in the skies.

(Here, I think, Derzhavin miscalculates a little. The first image of the sparks-suns turns out to be more expressive and convincing than the second. To be sure, stars that twinkle endlessly all around, like icy particles of dust in frosty air, present an inspiring picture. But the problem is that, to the naked eye, particles of dust and stars look very much alike. For this reason the poet didn't achieve a clear-cut interaction between both levels, a sharpness of comparison, but instead the image as a whole does not assume the cosmic proportions found in the first case, when sparks fly in all directions like suns that have just been born, and the suns fly in all directions like sparks.)

In the following stanza a comparison is built, not on scale or quantity, but on the intensity of light, of radiance. Structural continuity exists here as well: quantity, spatiality, degree.

> Millions of ignited heavenly bodies
> Flow into the immeasurable;
> They fulfill Thy laws,
> Emitting life-giving rays.
> But are they fiery icon-lamps
> Or a mass of glowing crystals, . . .

In other words, heavenly bodies that have cooled and turned to ice, but nevertheless sparkle, albeit with a cold light?

> Or a boiling multitude of golden waves, . . .

In other words, a molten mass of sun with its spray, tempests, and protuberances?

> Or the burning masses of ether, . . .

In other words. . . . I don't even know what the wise man of the eighteenth century was picturing in his mind's eye:

> Or all the glowing worlds together
> Before Thee—are like night before day.

We must not forget that "I" and "Thou" are contrasted. Consequently, the mightier and grander the "Thou," the smaller and more humble becomes the

"I." This juxtaposition reaches its apogee in the poem's fifth stanza:

> Like a drop released into the sea
> Is all the firmament before Thee.

Several contemporary poets often call the earth the "firmament" by mistake. They know that the word "firmament" exists. The earth's surface is firm, and herein lies the confusion. In fact, in Old Slavic "firmament" included the sky along with the entire expanse of the universe surrounding the earth.

How large is a drop if it is released into the sea? This means that we must imagine the entire expanse of the universe (and it is boundless, as Derzhavin states in this ode) as a drop and then release this drop into the sea, in order to understand. . . . But if the entire unimaginable universe is just a drop, then of what significance are the expanses perceived by the human eye? And of what significance, then, is a mere person—one hundred seventy centimeters in height and weighing eighty kilograms?

The concentric circles sharply decrease in size. Everything relates to the insignificant, not even microscopic, point—namely, to a nonentity.

> Like a drop released into the sea
> Is all the firmament before Thee;
> But how great is the universe I see?
> And what am I before Thee?

Since this last juxtaposition seemed inadequate to the poet, he expanded it:

> In this ethereal ocean,
> Worlds are multiplied by a million

(We must yet multiply the boundless cosmos by a million)

> A hundred such worlds—and this,

(Plus another hundred such cosmoses)

> When I dare compare it with Thee,
> Becomes a mere speck,
> While I before Thee—am naught.

The last word has been uttered. Indeed. If we multiply something that is boundless by a million, and then take a hundred such boundless worlds, and all of this is only one speck, then the real speck becomes even smaller than nothing.

But starting from this point the concentric circles of the composition begin to expand. The compressed spring unwinds in the opposite direction, to its limits. This lyrical daring—a *coup de metre*—is a stroke of genius, that's what the seventh stanza of Derzhavin's ode is!

> Naught!—But Thou shinest in me
> With Thy great goodness,
> Reflecting Thyself in me,
> Like the sun in the smallest drop of water.
> Naught!—But I feel alive,
>
> With a certain hunger I fly
> Always soaring upward;
> To be with Thee is my soul's hope,
> To ponder, grasp, and comprehend:
> I am—and, verily, so art Thou!

The daring stroke of a sniper. Although thus far not much has been achieved from the standpoint of the "unwinding of the spring" and the expansion of the concentric circles. Only existence itself, the very presence of the "I," has been affirmed. Indeed, I feel alive, I live, and think; this means that I really exist, no matter how insignificant I am. And furthermore (in the spirit of how things were understood in Derzhavin's idealistic time), insofar as a reflection in a mirror exists, then what is reflected in the mirror also exists, even if the mirror itself (or, say, a droplet of water) is infinitely small. But Derzhavin goes even further. He speaks of the reverse dependence of the two entities he contrasts: "I am—and, verily, so art Thou!"

But what about the vast scale? It didn't occur to the poet to hurry. Since the decrease in size happened gradually, the increase in size would happen gradually as well. The main thing is that the "I" has already been introduced. Notice how the first tentative little step is taken:

> Thou art!—nature's ranks proclaim,
> To me my heart repeats the same,
> My reason, too, convinces me:
> Thou art!—then naught am I no more!

Yes, the first little step, the first movement from out of the dust, out of the depths, out of nothingness. Where will this movement lead?

17

I am part of the whole universe, . . .

A small particle, but of what a universe in return! Does not its grandeur shine its rays even on me, an insignificant particle; does not the particle derive warmth from the knowledge of its belonging to something grand and even immeasurable? But the movement upward out of the dust quickens:

> I am part of the whole universe,
> Placed, it seems, in the honored
> Middle of all of nature,
> Where Thou created corporeal creatures,
> Where Thou began Thy heavenly spirits
> And linked the chain of being with me.

Like a boulder that breaks loose and brings on a mountain rock-slide, so this thought, taking form, immediately brings down an avalanche:

> I am linked with worlds in the cosmos
> I am the last stage of matter,
> I am the focus of all that lives,
> A trace of all that is divine,
> My body will be reduced to dust,
> I command thoughts like thunder,
> A king—a slave, a worm—a god!
> But if I am so marvelous
> Whence did I come?—who knows,
> But by myself I could not be.[20]

We'll forgive Derzhavin. After all, Darwin didn't exist in his time and the poet couldn't suspect that hairy monkeys would be to blame for everything. If he had known this, it is possible that not only would he not have written his ode, but he would not even have wanted to live; but, thank goodness, he didn't know any of this yet and the ode was written.[21] One of the best Russian philosophical lyric poems. And, personally speaking, I can't imagine anything that could be compared with it. . . .

Meanwhile, we continued on our way, and soon our *Raketa* began to seek a place where it could edge closer to the bank. Here (to our left, and we were coming from Novgorod) the riverbank was rather steep and high. Immediately the esteemed academicians, writers, and artists, somehow having become more youthful and consequently more alike, in a friendly and boisterous

18

throng poured out of the *Raketa* and, racing with one another, began to scramble up the steep, green banks. However, they didn't all scramble to the top at the same rate: their hearts were worn out to varying degrees. One who was short-winded stopped to catch his breath, while another threw up his hands altogether and clambered back down to the water, where he began to search for a more gently sloping path. But the fact of the matter was that there were not and could not be any paths here. . . . The merriment of those who had managed to scramble up the steep bank subsided and died away by itself. The entire surrounding landscape suddenly presented an unbelievable sight.

Many places have been visited by war. Even in the environs of Moscow here and there one stumbles upon (encounters while skiing in the woods) an old trench or the modest obelisque of a common grave. But the ensuing years have wiped away the steel rubbish of war, have erased all the traces, and now, even in Stalingrad or in the Kursk Bulge, there remains only the earth, possibly laden with the admixture of shrapnel, but even so, on the surface everything has been cleaned up; there are no reminders of past battles.

This place was unique, if only in the sense that after the fighting moved on to the west, no one returned here anymore. Probably immediately afterward they hurriedly cleaned things up and cleared away the largest, most obtrusive items: the occasional cannons, machine guns, and weapons in general. And the corpses that had not been covered up with dirt. Then they didn't return to this place anymore.

The first object I tripped over was a Soviet helmet pierced by a bullet or a piece of shrapnel.[22] The land was slashed by trenches and dugouts. Everything was crumbling to pieces and overgrown with grass and bushes. If you poked at the ground (and you didn't even have to poke) you could see that everywhere lay fired cartridge-cases, helmets, some kind of wheels, torn pieces of steel, bones and skulls.

The academicians fell silent and the artists were subdued, while the writer holding the pierced helmet grew thoughtful.

Nothing remained of the estate Zvanka, or even of its layout, here on the lofty bank of the Volkhov. Everything here had been irrevocably destroyed, erased from the face of the earth. But, even so, let us overcome a certain inertia, come to our senses after this first impression, and consider these conclusions:

1. Derzhavin was a great Russian poet, or at any rate our greatest poet of the eighteenth century.

2. Have there been cases in which dilapidated memorial structures were rebuilt from the ground up? Yes, there have. Take even Pushkin's house at Mikhailovskoe. Recently, on Sadovoe Circle,[23] Griboedov's[24] house, which had been torn down not long before, was rebuilt down to the last brick. They

came to their senses, and they rebuilt it. An oft-cited example is the so-called Old City in Warsaw—Staro miasto—built up anew after the war. Now, next to Staro miasto, the Poles are reconstructing King's (or Warsaw) Palace, down to the last brick. Examples, then, can be found.

3. Is it too much of a luxury for a government like ours to rebuild an eighteenth-century estate, and specifically Derzhavin's Zvanka, from the ground up? No, it's not such a great extravagance at all. Is it really impossible to decide in favor of such an expenditure?

4. Will this expenditure be justified materially in the near future? It will. Novgorod is a city for tourists, including foreigners. An additional tourist attraction would be the estate of one of Catherine's magnates, its daily routine, its setting (also taking into account that this nobleman was Derzhavin), and a one-and-a-half-hour walk along the Volkhov; all this, indisputably, would add up to a profitable tourist venture.

5. Will the reconstruction of Derzhavin's Zvanka be justified in other, nonmaterial ways? It will justify itself through the centuries, as long as the Russian language continues to exist.

Moreover, it would even be possible to reconstruct the external structure of Khutynsk Monastery, where Derzhavin was buried. As one travels by river toward Zvanka, this monastery is located on the right. It would make the excursion more picturesque. The boat could dock at the monastery (additional profits from the tourists), and one could lay at the poet's grave the flowers that are his due.

Immediately after the war, when Derzhavin's grave was located among ruins, the most expedient measures were taken. Rather than reconstructing the monastery complex from its ruins, they transferred the poet's remains to Novgorod, where they lie to this day in the Novgorod Detinets (kremlin) beneath a modest obelisque and a plaque bearing the inscription: "Gavriil Romanovich Derzhavin, Actual Privy Councillor and Bearer of Many Decorations." The plaque is old, having been transferred from the previous grave. I found it amusing when I saw it for the first time. As a matter of fact, Suvorov asked Derzhavin for a beautiful and triumphant (one must presume) epitaph for his future gravestone. "Here lies Suvorov," the poet answered without delay. "Lord, have mercy, that's good!" agreed the celebrated general.

One can understand the difference between these epitaphs, however. Suvorov possessed many titles and decorations, but we all know that there were many of them. Just as we know that Derzhavin was a great poet. But not everyone knows that he was an "Actual Privy Councillor and Bearer of Many Decorations."

We returned to Novgorod toward evening. And it seemed that no one was dissatisfied that, instead of taking a pointless and mindless ride around Lake

Ilmen, we had visited Zvanka, which until several hours before had been for most of us, if not an empty-sounding name, then something vaguely remembered from our school days long ago.

AKSAKOV'S TERRITORY

AT THE BEGINNING of this essay I should have written, in the spirit of Aksakov's unhurried, thorough, and meticulous manner, "Opening Remarks." I remember that this is precisely how several chapters in Aksakov's *Notes of a Rifle Hunter of Orenburg Province* begin: "Opening Remarks to the Description of Gamebirds of the Swamps," "Opening Remarks to the Description of Waterfowl." And how good it sounds! Opening remarks. To explain at the outset, why and for what purpose, and where the action will take place, and what kind of action it is, and how it began, in what kind of setting, and what preceded it; and then the web will be untangled in an orderly manner. And "opening remarks" are tantamount to finding a loose end in a ball of yarn. After all, each undertaking must start from somewhere, even if we take these notes[1] as an example. The term "opening remarks" also establishes a rhythm and imposes an unhurried quality—or at any rate, continuity—on the story. After having written "opening remarks," one cannot race through everything afterward, jumping from one subject to another.

I could also have begun the essay in this way: "I left my grandfather, I left my grandmother. . . ."[2] For the first ten-day period of December one of the southern republics with a coast on the Caspian Sea planned a huge literary festival. They invited about one hundred and fifty guests from Moscow, the other republics of the Union, and other countries. Through the overcast days in and around Moscow a blue sky appeared, the kind of hot weather one doesn't get in these parts even in the heart of summer, bright flowers (roses and gladioli), striking faces, motorcades from the airport to the hotel, from the hotel to the opera house, and later to various regions of the republic where once again would be—striking faces, clothing, and flowers and speeches, not to mention the festive meals, also with bright colors and distinct aromas.

As I was told later on, everything was just like this. And it even exceeded expectations. The writers who had traveled to this festival were transported, not in buses, but distributed among cars, "Volgas," in groups of three or four. According to some accounts, each writer was allotted a car. In the hotel rooms various bottled drinks were always available on the tables and were constantly restocked. Without charge. Fragrant shashlyk, chicken cooked with bacon, streams of pomegranate juice, succulent greens.

"Well, did you travel around the republic?"

"Are you kidding? Of course! You want to know how we traveled? We would drive about thirty kilometers from our starting point, and the inhabitants of the next village would be standing by the road: Young Pioneers[3] holding bouquets of flowers and gray-haired old men with trays in their hands. Bread and salt. Grapes and other kinds of fruit. Handshakes, embraces, welcoming speeches. Poems and songs are heard. Smiles glow. The clear sky glows. The red sun glows. Automobiles glow in the sunlight. And then in the evenings, large-scale meetings with workers, collective farmers, oil industry workers, fishermen. . . . No, you should have gone with us."

Yes, I didn't go. "I left my grandfather, I left my grandmother. . . ." Since right before this there had been another, similar temptation in a republic even farther away, but no less vibrant, with snow-white mountain peaks and green pastures, slow-moving glaciers and rapid rivers. And a bit later yet another trip hung over me, however strange it may seem, even more distinctive, which was linked with the centennial of an eastern classical writer.

I know almost for certain that many of my brothers-of-the-pen like very much to take part in such excursions. And one should take part in them, one should strengthen friendly, collegial relationships. I even took part by not sparing my stomach, but for several years now have slacked off and have applied the brakes in the face of sand in the hourglass that keeps diminishing at a faster rate. No matter what one does, the sand keeps on flowing.

Perhaps at this point I've adopted a tone that is too elevated and pathetic. Perhaps simply with age one begins to feel that for the soul it's somehow more gratifying to spend a week sitting at one's desk, looking out the window from time to time at a quiet, overcast day in the area around Moscow, than to listen to the distinct, loud applause of festive occasions. To put it more plainly, it must be that I've grown tired. Amuse yourselves, my friends, without me. To console myself I have the children's tale: "I left my grandfather, I left my grandmother. . . ."

How could I not start laughing, having in the course of the past two years

refused to travel to Tashkent and Frunze, Alma-Ata and Baku, Tbilisi and Erevan, when from the telephone receiver I heard the sweet voice of the employee of *Literaturnaya gazeta* and realized that they were offering to send me to the
Orenburg region for a week, to the little town of Buguruslan?

"What?! To Bu-gu-rus-lan?"

"Yes, to Buguruslan," confirmed the same sweet voice of the employee,
whose name was Tanya. "Would you like me to send you the letter I'm holding in my hands?"

"First, perhaps you could tell me anyway what I haven't seen in this . . .
Buguruslan."

"All right. You haven't seen there what condition the former estate of
Aksakov is in. And what is being done to preserve it. And what can be done
about it."

Tanya allowed me to keep my silence to my heart's content, and then stated, no longer with a questioning intonation, "So I'll send you the letter. . . . "

. . . The employee Tanya, of course, couldn't know entirely what Aksakov
means to me, both as a Russian and as a professional writer. Because we ourselves cannot assess either the influence this or that literary phenomenon has
exerted on us or its total significance for the organization of our consciousness. And a measure for it doesn't even exist. One person's influence on the
soul of another cannot be calculated in centimeters, grams, or even (a more
attractive and appropriate word) in *zolotniks*.[4]

At first there exists for us only the word "Aksakov." Maybe for some people
it continues to exists only as a word—not an empty one, incidentally, but one
containing in itself a completely specific phenomenon and a concrete meaning. This is often the case with popular words. One might not know the particulars, for example, of what Sedov discovered or where he traveled. But as
soon as someone mentions the name "Sedov," a person will immediately say:
"Yes, everyone knows him. The Arctic. The Arctic Ocean." In contrast, upon
mentioning Aksakov no one would conjure up images of polar bears and ice
hummocks. Aksakov? What do you mean—Aksakov? . . . That's our Russian,
native environment. Due to a confluence of circumstances, a person might not
have read anything by Aksakov; but, even so, the impression of him as a poet
of our native environment exists in that person's mind unwittingly somehow,
originating from an unknown source, as if he or she had entered this world
with an impression that was fully formed.

That's how it was at first with me as well. During the first fifteen, prewar
years of my life, living in the country and attending secondary school, I somehow never got around to reading Aksakov. It's possible that the small school
library didn't contain any of his works. And there was also the order of prior-
ities. One had to read Pushkin along with Lermontov, Gogol with Turgenev,

Nekrasov with Goncharov, at the very least Oblomov's dream, within the bounds of the school's program; and then, in addition, one had to read what the school persistently recommended: *Bruski: A Story of Peasant Life in Soviet Russia* by Panfyorov, *The Red Landing Force* and *Chapaev* by Furmanov, *The Iron Flood* by Serafimovich, *The Rout* by Fadeev. . . . And also (this time without the school's recommendation) *The Three Musketeers, The Castaways, The Hunchback of Notre Dame, The Man Who Laughs, The Mysterious Island, Twenty Thousand Leagues Under the Sea,* "The Headless Horseman,"[5] *The Last of the Mohicans.* . . . And also all of Jack London's works, and also *War of the Worlds* and *The Stranger.* . . . No, for some reason it turned out that during the first fifteen years of my life Aksakov's turn never came. But I might add—there's a good reason, a circumstance that mitigates any guilt on my part: we didn't study him in school. And didn't even mention him. But meanwhile the word already existed: "Aksakov." Where did it come from? It's a riddle and a mystery. From out of the air. It originated spontaneously in my consciousness. But I testify that it was already there. I hadn't read him yet, but I knew that Aksakov existed in the world.

Zharov and Bezymensky were included in the school's program. But Aksakov, I repeat, was not in it. But what does the school have to do with it? After all, there is literary scholarship. The cultural process. MGU.[6] Pedagogic institutes of higher learning with departments of literature. How many critics and scholars! And so, S. Mashinsky, the contemporary Aksakov scholar and author of a monograph about him, emphasizes (I quote from his foreword to the four-volume collected works of S. T. Aksakov, 1955):

> It would be difficult to name another major writer of the nineteenth century whose creative work has been studied so insufficiently. His literary legacy has not been compiled; and many of his works, lost in journals and newspapers contemporary to him, have not been brought to light. In addition, those works not published during the author's lifetime and to this day remaining the property of archives have not been taken into account.
>
> S. T. Aksakov is a difficult writer to comprehend. Moreover, the tradition of studying his work doesn't exist. For a long time his name has not attracted the attention of scholars. During the years of the Soviet Union's existence not a single book about him, not even a pamphlet, has been published.

The paradox: for a long time the name of Aksakov has not attracted the attention of scholars, but it exists in the consciousness of the people. Let's be serious, did it attract any attention? And if it did, why didn't scholars study his work? It couldn't be the difficulty of comprehension, as Mashinsky indicates, that frightened away prospective scholars. We do acknowledge, however, that no matter how dear, how priceless Aksakov is to us, he's still not Shakespeare, or Goethe, or Balzac, or Hamsun. *Notes on Fishing, Notes of a Rifle Hunter of Oren-burg Province, A Family Chronicle,* and *The Childhood Years of Bagrov the Grandson*—can it really be possible that these works are so difficult to comprehend that not a single literary scholar has dared to approach them and interpret them?

Well, there you are. In the program at the Gorky Literary Institute under the auspices of the Writers' Union of the USSR, Aksakov was also not covered. But he existed, even though at that time no one was reading his works. And it's not only because during their walks students could wander to Abramtsevo—that is, the former small estate of Aksakov, which subsequently became the property of Mamontov and somehow became linked memorially with some of the most important figures of our native culture: Vasnetsov, Vrubel, Serov, Korovin, Polenov, and Nesterov. And it's not only because our institute stood side by side with the Kamernyi Theater (now the Pushkin Theater) and once, when Professor Galitsky (who taught Western literature) fell ill and two hours of daytime opened up for me, out of some kind of intuition I decided to go to the morning children's show (fifteen steps from the gates of the Institute to the entrance doors of the theater); that morning they were performing *The Little Scarlet Flower* by Aksakov. Fifteen steps—do they not constitute a single step? That's just it: I took a single step and, to put it more plainly, stepped from the Tver Boulevard, from the asphalt sidewalk, from a Moscow street of 1946, from our dormitory in the basement, from oil stoves and bread-ration cards, from pearl-barley and powdered *kisel',*[7] from streams of automobiles and stuffy shashlyk-houses—I stepped from everything that characterized the reality of Moscow in 1946—and, as if by magic, found myself in an entirely different world. In the tale the beautiful girl was required to put a small golden ring on the little finger of her right hand in order to be transported from place to place at the other end of the world, but I didn't need a small ring. The grayish paper of the theater ticket that cost ten rubles—one ruble in today's terms—consti-tuted the entire miracle-working talisman that immediately transported me to another time and another world.

But I'm not concerned with a theater performance at the moment. After all, this was not the first time I had been to the theater. And we also made the rounds of movie-houses, where one could also be transported for a time to the Devil knows where. I have in mind instead the atmosphere of the tale, that affectionate Russian quality in which I immersed myself and which immedi-ately gave rise in my soul to kind and radiant feelings. So sharp was the con-trast, so great the distance from atmosphere to atmosphere, from climate to climate, from feelings to feelings, that I sat without breathing, enchanted, astonished, and dumbfounded, sensing that something warm and radiant was washing over my soul.

At the moment, in perusing for my essay various materials associated with the places Aksakov had known, I came across a brief review in the newspaper *Yuzhnyi Ural* from 1963, entitled "In the Village Where *The Little Scarlet Flower* Bloomed":

> Long ago in a tall log cabin that stood in the center of a hamlet on the bank of the Buguruslan River, the seven-year-old boy Seryozha[8] Aksakov tossed and turned with a fever and suffered from insomnia. In order to help the sick child fall asleep they called the housekeeper[9] Pelageya, who was known in the neighborhood as a talented storyteller.
>
> "And Pelageya came," S. T. Aksakov related many years later, "a stout woman who was no longer young; she sat down by the stove, rested her head sorrowfully on one hand, and began to speak in a slightly singsong voice, 'In a certain kingdom, in a certain country. . . . '" And a marvelous tale was born, which was later reworked artistically by the writer Aksakov. It is known today to Soviet children by the title *The Little Scarlet Flower*.
>
> Not "in a certain kingdom," but in the homeland of the housekeeper Pelageya storytelling is carried out these days by compatriots of Aksakov—the collective farm workers of the agricultural artel Rodina [Homeland—V. N.]. A session lasted for several days. . . .

But stop. We'll read the end of the brief review a bit later, when we reach the appropriate place in this essay. We'll save the end of the review, for we'll be able to make use of it later. But for now let's marvel at how, in the nineteenth-century backwoods of Buguruslan—in a real log cabin, where the senior Aksakov (the grandfather) was half-literate and despotic, and where the grandmother languished after his death because "there's no longer anyone to be afraid of now"—how in that place there miraculously arose and bloomed a little scarlet flower, if we keep in mind as well the tale itself, and if we also keep in mind the talent that was born in the soul of Seryozha Aksakov. From out of low-ceilinged and cramped peasant *izbas*[10] covered to their very roofs by snow from fierce snowstorms, away from bread and *kvas*,[11] from the labor of the peasants and shirts of sackcloth, what power of imagination was needed in order to transport that same housekeeper Pelageya to kingdoms across the ocean, to palaces of white marble and gardens with glorious flowers in bloom, to fragrant victuals and harmonious music, to large, well-rounded pearls and to sparkling semiprecious stones, to pearly fountains, and to lace clothing. . . .

But if we stop and think about it, all these things are attributes of and constitute the entourage, the necessary, usual accessories for many fairy tales. And what is most important in a fairy tale? That's an easy question to answer. Most important in a fairy tale are—goodness and love. And the fact that bad feelings— envy, greed, and egoism—do not triumph, while the darkest evil is defeated.

Defeated by what? By love. By goodness. By gratitude. So you see, to find these qualities, neither the housekeeper Pelageya, nor her seven-year-old listener Seryozha, nor the Russian people in general, had to travel for miles and miles to another kingdom or country. These qualities exist in a person's soul; they are the essence of the soul and its best inspiration. They in particular comprise the little scarlet flower that is planted in each person's soul; it matters only that the flower grow and bloom.

Perhaps I myself am a proponent of vengeance, a proponent of direct force in the struggle against the forces of evil. Perhaps I myself might have cried out, in the place of Alyosha Karamazov, "Shoot him!" And perhaps the Goths were even right, when they were being baptized in water, not to immerse their right hands with swords in them, but instead to raise them over their heads, so that these right hands would remain weapons of power and vengeance. Perhaps even they were right, who during our childhood, in place of verses appealing to the positive movements of the soul—such as, for instance, the following:

> In the evening stars were shining,
> Frost was crackling all around.
> Through the village walked a young boy,
> Body shiv'ring, bluish skin.
>
> Babushka walked through that village,
> Saw the orphan poor and cold,
> Took him in and made him warm,
> Gave him something good to eat. . . .[12]

—perhaps they really did the right thing when, in place of these verses, they taught us to recite from memory: "Airplanes are flying, machine guns are banging," and likewise the little songs in which appeared machine-gun carts, grenades, the sailor Zheleznyak, young drummers, sabers, steeds, and later, even tanks and tank-operators. . . .

But I can't forget that moment (those two hours) when by chance I entered the theater for a morning performance and experienced the warm and radiant feelings that washed over my soul, and saw the triumph of the simple, but most precious, of course, human qualities—goodness, gratitude, and love.

Aksakov himself wrote the following to his son Ivan:

These days I'm preoccupied with an episode in my book: I'm writing a fairy tale which I knew by heart as a child, and which I would relate to everyone for fun, with all of the humorous embellishments of the storyteller Pelageya. It goes with-

29

out saying that I had completely forgotten about her; but recently, while rum-
maging through the storeroom of my childhood memories, I found among many
different kinds of junk a pile of fragments from this tale; and since it will become
part of *Grandfather's Stories*, I've set about reconstructing it.

"Among many different kinds of junk" he found, not just a pile of fragments
from an old fairy tale, but a precious diamond and, more precisely—a living,
enchanting little scarlet flower; and still, over one hundred years later, the
mere contact with it, as with our own "junk" (primarily a pile of debris), is
enough to enable us to discover suddenly the living, precious sprouts that are
even capable of blooming, if we keep them warm and water them and create
the appropriate climate for them.

Subsequently I went to see *The Little Scarlet Flower* more than once at the
Pushkin Theater. And one time I met in the auditorium—whom would you
expect?—an actor of this same theater, the wonderful Russian Soviet actor
Boris Petrovich Chirkov. And since he and I (to adapt my style to Aksakov's)
were somewhat acquainted, I expressed my amazement.

"And you'll come to see this play many more times, regardless of your age.
Perhaps you'll even come sometime when you're an old man. Every age can
find its own delight in this tale. It's an immortal, beautiful tale. And most
important is the goodness in it."

The same wonder-working qualities emerge as well in the remaining (main)
works of Sergei Timofeevich Aksakov. Do I need to say that, in that same year,
as I was leaving for my vacation in the country, I grabbed the heavy volume of
this writer's works to take with me? And it was at this time that my long-await-
ed, genuine encounter with him took place.

Aksakov's roots go back to the eighteenth century. And it's not surprising.
He managed to live for eight years in the eighteenth century (he was born in
1791); one must bear in mind that the old century, in preserving its inertia, did
not immediately, in just one year, relinquish its atmosphere to the newly born
nineteenth century. And one must also bear in mind that the atmosphere of a
new era did not spread at once from both capitals to Samara, Ufa, and the
Orenburg steppes.

Something else is amazing—namely, that Aksakov, with his roots (and lexicon) reaching back to the eighteenth century, reaches us with his branches;
and that, in touching these branches, we see that they don't belong to some
sort of lifeless debris and brush, but to living, full-blooded branches.

In associating with professional writers—that is, at least with qualified readers—one is convinced on the basis of their experience, if one doesn't trust solely one's own, that nowadays not everything can be reread easily and with pleasure, let alone with real delight. If for the twentieth time one can, with one's

original enthusiasm, reread *The Captain's Daughter,* or Pushkin's prose in general, including even "A Journey to Arzrum" or *A Hero of Our Time,* then one can reread *Yury Miloslavsky* or *Prince Serebryany* only by disciplining oneself somewhat. It's a terrible thought, but now one rereads most of Turgenev's works without lively and expectant enthusiasm, while the enlightened novels by Chernyshevsky for the life of me no longer attract me. . . .

But why name the classics? Recently I wanted to reread *Bruski: A Story of Peasant Life in Soviet Russia,* as well as some other related prose contemporary to that novel, but my efforts proved disastrous. What could this mean—that these are long departed coryphaei? Recently, out of interest, I picked up several novels by writers still living. . . . All right, all right, that's not the subject of my essay. I only wanted to say that Aksakov and his almost pre-Pushkinian lexicon can be read today with as much enthusiasm, preserving for us all of the same enchantment and fascination as was the case one hundred years ago. There's some kind of secret here, some kind of riddle, the incomprehensible sorcery of the artist; there's something bewitching here that one cannot explain, not by any form of psychological analysis.

It is well known that young people nowadays have become proficient in speed-reading. My daughter reads eight times faster than I. I also have a friend who is used to reading what is known as "diagonally" and is still able to grasp the basic content and information contained in the pages she reads by this method. I ask her several times, question her—she's grasped it. And now I give her Aksakov to read, as an experiment. We won't reproach the young reader for not having read Aksakov yet. As I remember, even I myself when I was a student. . . . And many people, like I myself at that age, know what Aksakov represents and understand what place he occupies in Russian literature, but to read. . . . We won't judge them severely. So, as an experiment I give Aksakov to the reader who is used to reading quickly, diagonally, swallowing it whole, grasping the informational aspect of the text. I observe: right now a page will be turned, then another. Speed-reading. The twentieth century. No, the page is not being turned. The page is being read slowly, with the reading of each line, each word. Only from time to time exclamations are heard:

> "How charming!"
> "How enchanting!"
> "Heavens, it's so good!"
> "Listen, you know, he's a real magician, a sorcerer!"

But let's digress a little from the artistic merits of Aksakov's prose and first elucidate for ourselves, in amazement, that a phenomenon as precious and unique as Aksakov could very easily not have existed in our native literature. I have in mind, not the extremes (the boy Seryozha might not have been born, might have died in childhood, might have drowned as a twenty-year-old during the trip across the Volga, etc.), but rather the simple circumstance that, until he was fifty-four years of age, this person in fact had not been a writer. More specifically, had not been "Aksakov." And, even more specifically, had been both Aksakov and a writer, but until he reached the age of fifty-four had not written anything "Aksakovian"—that is, what is understood when one utters the name Aksakov. And to wait until the age of fifty-four is taking a risk, isn't it? Many of those who became well known, and even great, writers died at a much earlier age than that up to which Sergei Timofeevich, so to speak, had tested fate and its patience.

To be sure, the brief sketch "The Snowstorm" was published anonymously in the almanac *Dennitsa* in 1834. To be sure, commentators say that "The Snowstorm" seemed to herald the emergence in him (Aksakov) of a major artist-realist and paved a clear way to *A Family Chronicle* and *The Childhood Years of Bagrov the Grandson.* To be sure, according to S. Mashinsky: "It was Pushkin who first recognized the talent in this description (of the snowstorm). Testimony to this is that his famous depiction of the snowstorm in the second chapter of *The Captain's Daughter* undoubtedly was inspired by Aksakov. A mere comparison of the corresponding passages from both works is enough to enable us to feel just how close Pushkin was—even in the textual aspect—to the author of 'The Snowstorm.' "[13]

Maybe all of this is true; but, first, "The Snowstorm" appeared in 1834 when Aksakov was already forty-three years old; second, it is easy to say now that "The Snowstorm" "heralded a birth and paved a clear way," but his basic works, which emerged ten to fifteen years later, could easily not have been written, as we shall see below, and then "The Snowstorm," as the saying goes, would have been left hanging in the air, and would have heralded absolutely nothing at all, and would not have paved a clear way to anything at all. Third, this brief sketch was written by chance in 1834, at the request and under the pressure of Maximenko, for the next issue of the almanac, and it was an exception to the work of Aksakov himself, who considered his primary occupation to be that of writing theater reviews and corresponding with friends, albeit such friends as Gogol.

So he wrote down a vivid picture of nature in Orenburg; well, what of it? It's just a trifle. He even published it without his signature. Fourth, what does "The Snowstorm" actually consist of? Nine little pages of typographical text; let's face it, it's not *Woe from Wit* or "Princess Mary" or "Taman." "A study," as we would say today—a sketch that for twelve whole years afterward was not supported by a single line of literary work by the writer.

In a few words, Aksakov's path to his "Aksakovian" prose was as follows.

We'll start from the place where a boy was dying from a lingering illness that lasted one and one half years.

> Sometimes I would lie in a state of oblivion, in a state somewhere between that of a dream and a faint: my pulse had almost stopped beating, and my breathing was so faint that they would place a mirror up to my lips in order to determine whether or not I was still alive. But I remember many things that were done to me at that time and what people said in my presence, assuming that I no longer could see, hear, or understand anything—that I was dying. The doctors and everyone around me had long since condemned me to death.

The relative Cheprunova had advised the boy's mother:

> "Even the doctors, and the priest, too, have told you that he doesn't have long to live. Resign yourself to God's will: lay the child down under the icons, light a candle, and let his little angelic soul leave his body peacefully. After all, you're only interfering with it and upsetting it, and you can't be of any help. . . . " But my mother would become enraged at such words, and would answer that as long as a spark of life glowed inside me, she would continue doing everything she could in order to save me.

They were driving the boy somewhere, and on the road he began to feel much worse:

> I began to feel so ill and grew so weak that we had to stop; they carried me out of the carriage, made a bed for me in the tall grass of the forest's clearing, in the shade of the trees, and as they laid me down my body was almost lifeless. I saw and understood almost everything that they were doing around me. I heard my father crying and comforting my despairing mother, and I heard her praying fervently as she lifted her arms to the heavens. I heard and saw everything distinctly, but I couldn't say a single word, couldn't move a muscle. And suddenly it was as if I had awakened and had begun to feel better, stronger than usual. I liked the forest, shade, flowers, and fragrant air so much that I asked them not to move me from this place. And we remained here this way until evening. They unharnessed the horses and let them graze in the grass right near me, and this pleased me. Somewhere they found springwater; I heard them discussing it; they built a fire and drank tea. . . . I didn't sleep, but felt in unusually good spirits and experienced an inner contentment and peacefulness, or, more precisely, I didn't understand how I felt, but I felt good. . . . By the morning of the next day I similarly felt fresher and better than usual. . . . Hour by hour I felt better, and after several months I was almost healthy again; but this entire time, from the feeding in the forest's

clearing until my complete recovery, has almost completely been erased from my memory.

And so nature cured the hopelessly ill boy. Or, to be more cautious—in the lap of nature, in the clearing of a forest, in the shade of trees, where there were flowers and springwater, under the cover of a blue sky, the primordial stimulus for a cure took place, the healing process of a future, marvelous poet of nature began. I don't mean to say that Aksakov consequently, out of a feeling of gratitude or an unpaid debt, glorified nature; but evidently at that time her beauty so impressed the dying boy and filled his soul to such a degree, infused his being with such a sense of well-being, causing him to merge with her and lose himself in her to such a degree, that this feeling remained with him for the rest of his life as a heightened feeling for his native countryside, its quiet charm, its beauty, and its soul.

The child's talent, and later that of the youth, manifested itself in tangential, indirect ways. From the time when he was a child he liked very much to declaim and was able to read expressively. He retained this love and ability throughout his life. Incidentally, this declamation led to Aksakov's acquaintance with the already declining Derzhavin, for whom the young man would declaim and read for hours on end. He agitated the patriarch of his country's poetry with his expressive readings to the point that Derzhavin's doctors finally forbade Aksakov not only to declaim, but even to visit him.

All of this would be described later in detail under the title "An Acquaintance with Derzhavin." I recommend that you read it.

One could choose not to talk about this: just think—what a talent, reading someone else's poetry; but insofar as the writer of these lines also at one time started out by declaiming, by reading someone else's poetry at school festivals, and later at various amateur nights and propaganda-team gatherings, he can testify that such readings of others' poetry develops in a person two qualities that may never be needed if he or she never undertakes to write anything, but that may turn out to be extremely valuable as soon as this person picks up a pen and begins to write. I'll tell you about these qualities. First, a feeling for the word is ultimately developed. It is one thing simply to read a poem or an excerpt from a work of prose with the eyes, but another thing to learn it by heart and then test it, so to speak, many, many times on various listeners. One begins to understand how each word sounds, which word is in its place, and which one is not really in its place and is even entirely superfluous. The second extremely important quality for a future writer is the development of a sensitivity to the reader, a sensitivity to the auditorium, an internal, perhaps subconscious, sensitivity to what in particular is of interest to these readers, and what is not of interest to them. Do I even need to mention that when a writer is at work, he or she must maintain a mental image of a particular auditorium of listeners who will subsequently read his or her writing? So the quality of a sensitivity to the audience and of what will be of interest to it, and what will not be of interest, is extremely valuable for a writer; it is precisely this quality that developed in Seryozha Aksakov when he would often and unselfconsciously declaim the work of Derzhavin, Sumarokov, Khemnitser, Knyazhnin, and Dmitriev.

From the time of his childhood Aksakov possessed a talent for depicting other people, for imitating their conversations, and for mimicry. He was able to portray this or that relative so well that everyone would be rollicking with laughter. This talent gradually developed in him the artistic capabilities to which everyone who knew him would subsequently attest. He participated in many amateur performances (in particular, at the home of the famous Shishkov), and everyone regretted that his noble rank prevented him from becoming a professional actor. Because of this interest of his, Aksakov came to know and love the theater for the rest of his life, and throughout his life would write about it in journals, so that until his fifty-fourth year theater reviews constituted his primary literary activity.

From the time of his childhood Aksakov was also an outstanding storyteller. Whether he was relating to his mama how he and his father went fishing, or later telling Gogol about his childhood years, it all came out vividly and picturesquely, and it affected his listeners in an infectious way. After all, it was specifically with respect to his oral stories that Aksakov subsequently formulated a sort of writer's credo:

> Upon returning to Bagrovo I wasted no time in describing in detail everything that had taken place in Old Mertovshchino, first to my dear sister, and then to Auntie as well. In my liveliness and insurmountable, instinctive desire *to convey to others my impressions with the precision and clarity of an eyewitness in such a way that the listeners would receive the very same impression of the events that were being described as I myself had of them*—I began to mimic the crazy Ivan Borisovich's mumbling, grimaces, and bows.

(By the way, I once lifted out of one of Aksakov's texts the middle of this last sentence, specifically the words emphasized here, as the epigraph to *A Drop of Dew*,[14] and, when freed from the other details, they actually began to sound like a concisely formulated statement by Aksakov of his artistic method.)

And so, an insurmountable and instinctive desire to convey his impressions to others. . . . But we all know that this is the basic stimulus for a writer's work! This means that the writer's spirit had existed in Aksakov since his childhood and had simply waited for a very long time for its time to come.

While listening to Aksakov narrate stories aloud, Gogol more than once attempted to convince Sergei Timofeevich to take up the pen and write down his reminiscences. A literary document to this effect exists, namely, a letter Gogol sent to Aksakov from Ostende:

> It seems to me that if you were to dictate to someone your recollections of your past life and meetings with all the people you had occasion to encounter, along with faithful descriptions of their personalities, in this activity you would sweeten your last days enormously, and moreover would give your children many useful lessons in life, while all your compatriots would receive an exceptional understanding of what constitutes a Russian person. This is neither a trifle nor a feat of little importance in our time, when we need very badly to learn the real source of our nature. . . .

There also exists a statement by Yu. Samarin, who knew both Aksakov and Gogol well; he relates the following:

> I remember the very concerted attention with which Gogol, in fixing his eyes upon him, would listen, for evenings on end, to the stories of Sergei Timofeevich about the natural world on the left bank of the Volga, and about life in that region. He reveled in these stories, and in his face one could see such great pleasure that even he himself would not have been able to express it in words. Gogol confronted Sergei Timofeevich and demanded that he take up the pen to write down his recollections. At first Sergei Timofeevich didn't even want to hear of this, and even almost took offense; but later, little by little, Gogol succeeded in getting him excited about it.

The process of "getting Aksakov excited about it" continued for fifteen more years, and although the opinion exists that it was precisely Gogol who urged Aksakov on to authorship, in reality in Aksakov's life it took place much more simply and tragically. In our lives, incidentally, everything has turned out happily and well. But in the sixth decade of his years Aksakov began to experience a sharp decline in his health; specifically, he began to go blind. A damp, cold wind from the other world made itself felt. In order to enliven his last days somehow, Sergei Timofeevich decided to remember his golden childhood: the golden years spent in the open spaces of Orenburg, on the Buguruslan River, and in the hamlet of Aksakovo; his passionate preoccupation with a fishing pole and rifle; and the confluence with nature and all her benefits—he began to dictate what he remembered (he could no longer write because of his blindness), "to refresh [his] memory and for [his] own enjoyment," and in order to relive everything once more during his waking hours, to recall the unforgettable days. In other words, to put it crudely, Aksakov took up authorship in the strict sense of the word when, as the saying goes, the cock had already pecked him on the head.

And without this—well, what can one say. . . . Of course, he was Gogol's contemporary and friend. He occupied himself by writing theater reviews. He would have been familiar to specialists studying that time period, along with, say, Shevyryov, Pogodin, Shakhovskoi, Kokoshkin, Panaev, Samarin, Nadezhdin, Kavelin, Maximenko. . . . But Aksakov, of course—today's, our Aksakov—would not exist.

There are only four books which transformed Aksakov from a dilettante who wrote insignificant poems and theater reviews into a major Russian writer; which, immediately upon their successive publication, one after the other, attracted the general attention of both readers and writers; and which became part of our native literature's golden reserve. Moreover, two of them generally are not considered literary works. But nevertheless, they are literary works, and in a marvelous way: *Notes about Fishing*, 1848; *Notes of a Rifle Hunter of Orenburg Province*, 1852; *A Family Chronicle* and *Reminiscences*, 1856; and *The Childhood Years of Bagrov the Grandson*, 1858.

One notes that all four books were published in the course of a decade. In 1848 Aksakov was fifty-seven years old, in 1858 he was sixty-seven, and in 1859 he died.

And all of this during the last decade of a life that spanned about seventy years. This is what I had in mind when I mentioned earlier that such a valuable and unique phenomenon as Aksakov could very well not have existed in our native literature. The writer's first book, like all those that followed, was enormously successful, a reception he neither expected nor could have predicted. To be sure, he knew that he was not writing merely a technical manual for fishers and hunters. In a letter to Gogol he writes:

> I undertook to write a book about fishing not only in its technical aspects, but in its general relationship with nature as well; the passionate fisherman in me also loves the beauty of nature just as passionately; in a word, my work became a labor of love, and I hope that this book not only will be pleasant for the lover of fishing to read, but also for anyone whose heart is receptive to the impressions of an early morning, late evening, and luxurious midday. . . . Space in the book is devoted to the splendors of nature in the Orenburg region as I knew them some forty-five years ago. This occupation revived and refreshed me. . . .

Concerning the second book, which was written for his son Ivan (the one for whom a street is named in Sofia, the capital of Bulgaria), Aksakov writes:

It gives me pleasure to imagine myself starting to read these notes to you. Although I'm not satisfied with everything, I'm convinced that there is much that is worthwhile in them for the hunter with a soul, the naturalist, and the writer. I constantly try to keep myself from getting carried away by the descriptions of nature and other subjects tangential to hunting, but Konstantin and Vera emphatically insist that I give free rein to my will: your voice will decide this matter. Just as much as I fear fire, I dread the garrulousness of an old man who typically assumes that everything he knows no one has heard before and will be of interest to everyone.

If someone were to ask me to name the chief characteristic of Aksakov's books, and specifically a single one, my answer would be incomplete and imprecise, but I would identify it as "spiritual health." The spiritual health that involuntarily passes to the reader of these books infects (if one can speak this way about health) and enriches him or her. Vanity, nervousness, all forms of pettiness, haste, and the effects of being overworked (everything that prevents us from glancing around with calm, clear eyes) retreat to the background. Aksakov's unhurried narrative style, clarity of thought and feeling, purity and vividness of language—notwithstanding its seeming simplicity—and great expressiveness within its seeming unpretentiousness—all this underscores the fact that one reads his books with pleasure; and, in the same way that, in his early childhood, as we remember, virtually without a pulse or a breath, he suddenly felt the incomprehensible strength of nature entering him after they had laid him down on the grass under a tree in a forest's clearing, so the reader of these books, with shattered nerves and also virtually without a spiritual pulse, suddenly feels a surge of strength and the return of good spirits.

It is well known that in the 1860s, when Aksakov's books appeared one after the other, various tendencies and currents intersected with each other in Russian literature, especially since one tendency was represented by Chernyshevsky, Dobrolyubov, Herzen, and Nekrasov, while another—the right wing—was represented at the very least by the Slavophiles. It is noteworthy that in the assessment of Aksakov's books everyone was in agreement and united, expressing complete unanimity. It follows that Aksakov had touched, both in the Left and the Right, those deep and common nerves located deeper than differences of opinion that are no longer superficial but exist on a more profound level. I'll cite several passages from the foreword to his four-volume collected works, and also from S. Mashinsky's commentary, which contains some reviews of Aksakov's books by his contemporaries:

Our literature has never before produced a book like this one (Turgenev).

Such masterful descriptions, such love for those things he describes, and what extensive knowledge of the life of birds! S. Aksakov has immortalized them in his stories, and, to be sure, no Western literary tradition can boast of anything similar to the notes *Of a Rifle Hunter* [sic—V. N.] (Chernyshevsky).

The outstanding book by S. T. Aksakov, *Notes of a Rifle Hunter of Orenburg Province*, is the talk of all of Russia (Nekrasov).

None of our Russian writers can depict nature with such strong and fresh colors as Aksakov (Gogol).

There is nothing artful or complicated in his depictions of the landscape; he never flaunts anything or tries to show off. Even in her capriciousness nature emerges as kindhearted; as is the case with those possessing genuine and great talent, Aksakov never resorts to posturing before the countenance of nature (Turgenev).

Turgenev also noted that Aksakov's books on hunting enriched not only the literature of hunting but our general literature as well.

In writing about an excerpt from *A Family Chronicle*, Dobrolyubov stated: "The story doesn't contain any mannerisms or the desire on the part of the author to seduce the reader by means of artfully conceived action; what an amazing impression this little story produces." Subsequently, he devoted two well-known articles to Aksakov's work: "The Country Life of a Landowner from Days Gone By" and "The Various Works of Aksakov."

Herzen considered *A Family Chronicle* to be a book of enormous importance, while Shchedrin viewed it as a valuable contribution that had enriched our literature.

Turgenev wrote, "This is an authentic tone and style, this is Russian life, herein lie the directions of future novels." In comparing *A Family Chronicle* with *My Past and Thoughts*, Turgenev wrote the following to Herzen: ". . . these two books represent a truthful picture of Russian life, albeit from its two poles, and from two differing points of view. But our land is not only great and abundant, but spacious as well, embracing many things that might seem incompatible with one another."

Annenkov rightly wrote of the "sweet-sounding Russian language" of Aksakov.

Gorky's *nouvelle In the World* describes the significance of Aksakov's books for the course of the narrator's spiritual development. Along with several other books, they "cleansed the soul and cleaned off the scales from it."

According to Panaev, the most important feature of Aksakov's *Notes* is the profound, poetic feeling for nature that is typical of a great artist, along with the amazing simplicity of depiction that distinguishes genuine works of art.

This book contains so much simplicity, Panaev felt, that one can boldly exchange it for the dozens of so-called novels, *nouvelles*, and plays that in recent times have enjoyed rather noteworthy success in our country. There is more poetry in this modest booklet than one would find in entire volumes of various lyric and epic poems which have found acceptance here, and which do indeed possess some poetic virtues. Maybe it doesn't have the same significance for the specialist interested in fishing that it has for the artist and writer.

Nekrasov wrote to Turgenev that he was rereading Aksakov's book, and that once again he was delighted with it.

> This book, which the hunter will read from the first to the last page, benefiting from and enjoying it, will be read with enthusiasm as a novel by those who aren't hunters (*Trudy imperatorskogo Volnogo Ekonomicheskogo Obshchestva*).

> He does not attract our attention with stories about elephants and tigers, but focuses it so intensely on subjects that are known, to a greater or lesser degree, to all of us, that we are unable to tear ourselves away from his animated narration (*Zhurnal okhoty*).

> In the entire repertoire of the stage, with all its tragedies, conflicts, vaudevilles, and ballets, one won't enter the action with as much empathy as one feels for the fate of the feathered inhabitants of the swamps, steppes, and forests described by the author. This is what it means to know and love one's work: every small detail comes alive, and a simple description is elevated to the level of art (*Moskvityanin*).

> Those not yet acquainted with the work of S. T. Aksakov cannot imagine how very engaging it is, or the charming freshness that has been breathed into its pages. It's impossible to read this book without some kind of cheerful, clear, and complete sensation similar to the sensations that nature herself awakens in us: we cannot conceive of higher praise than this. . . . His style is extraordinarily appealing to me. This is genuine, Russian style: kindhearted and direct, versatile and deft. There is nothing pretentious and nothing superfluous, nothing forced and nothing dull in it—its freedom and precision of expression are equally remarkable. This book was written eagerly, and it is read in the same way (Turgenev).

> I heard [a reading of—V.N.] two remarkable literary works: *Reminiscences of Childhood* by S. T. Aksakov and "A Profitable Post" by Ostrovsky. I thought the first work was better than the best parts of *A Family Chronicle*. It doesn't have the concentrated, youthful force of poetry, but rather the sweet poetry of nature evenly spread throughout the whole (Tolstoi).

The critics wrote about the unparalleled unanimity of praise. Chernyshevsky and Dobrolyubov, Herzen and Saltykov-Shchedrin, Tolstoi and Turgenev—all of them considered Aksakov's book *A Family Chronicle* an event in Russian literature. Here are several typical lines from an unpublished letter of Shevchenko to Aksakov, dated January 4, 1858: "It's been a long time since I read your best-known work, but now I'm reading it once more, and with the same sheer pleasure that the tenderest of lovers experiences while reading a letter written by the dear one he idolizes. I thank you many, many times for this deep and heart-warming pleasure."

The words "magician" and "sorcerer," of course, don't explain anything. But at the same time, regarding Aksakov's prose, "rapture" is always accompanied by "amazement." And in reality the prose contains neither an engaging subject, nor a conflict between characters, nor either tragic or comic situations. Why, then, is it so interesting? Where does the pleasure in reading it come from? How does the author manage to touch the soul's most precious and profound chords? And, finally, why is his work enjoyed by people of the most diverse, sometimes extreme, biases and convictions? They (the Westernizers, for example) ought to curse and criticize it, but instead they are in ecstasy over it and delighted with it, along with their literary and journalistic, or to put it in stronger terms, public opponents.

It is impossible to explain Aksakov's secret, and he probably doesn't have one anyway. The fact that his prose is documentary still doesn't mean anything—aren't there lots of boring examples of documentary prose sketches in the world?

To be sure, all writers can be divided into those whose work is predominantly fictional and those who must be satisfied to describe what they personally have experienced as well as factual material. The first group easily creates and builds on complex plots, introduces a multitude of heroes and other characters (even if they sometimes originate from prototypes), conceives various interesting situations for these characters, causes various plot lines in the novel to intersect, brings together the heroes in sharp conflicts, and later resolves these conflicts.

The majority of writers fall into this first group. As a rule, their work consists of many volumes. Balzac and Dumas, Dickens, Romain Rolland, Gogol, Turgenev, Dostoevsky, Tolstoi, Jack London, Chekhov. . . . I cite the most obvious examples, and each of us could continue this list. The second group of writers is less inclusive, and its legacy is often limited to only a few small roots. In this group we can probably include Rousseau, Proust, Sterne, Saint-Exupèry, Hemingway, Herzen, Kuprin, Bunin, Prishvin, and Paustovsky. . . . Sergei Timofeevich Aksakov is also a shining example of this group:

> I'm not capable of replacing reality with fantasy. Several times I tried to write
> about imaginary events and imaginary people, but it turned out to be utter rub-

bish and seemed comical even to me.

> Those close to me have heard me say many times that my creative powers are not free-ranging; that I can write only while standing on the soil of reality, following the thread of an actual event; and that all my attempts to work in another mode turned out entirely unsatisfactorily and convinced me that I don't possess the gift of pure inventiveness.

In the meantime, literature (at least, Russian literature) increasingly began to gravitate specifically toward the second of the aforementioned forms. In 1854 the journal *Sovremennik* published the following words of an anonymous reviewer of a new excerpt from *A Family Chronicle*: "It is finally time, at least for the sake of experience, to change the form of today's belletristic works, to set aside love with its catastrophes and turn to the simplicity of a story, to add to it something ordinary or accessible and close to the actual flow of people's everyday lives."

Lev Tolstoi rightly noted that Russian artistic thought doesn't fit within the frame of a novel and is searching for a new form for itself. In another place the great artist expressed himself even more directly, in the sense that in the future writers will stop creating novels, but instead will write about life as it is in reality. This accurate thought, or more precisely, foresight, is impossible to comprehend in a simplified form. Not to invent plots, but rather to base one's writing on what one has experienced does not at all signify that one is controlled by the material and doesn't have the right to call oneself an artist. Have you ever heard an interesting book or film described by someone who has no talent for narration? "And he set out, and she says to him. . . . At this point the brother walks up to them, and the old man walked up to. . . . " After five minutes you can't follow anything, neither who he is, nor where she is, nor who walks up to them and says what. A genuinely interesting book is transformed into a long and boring recapitulation, in which everything eludes your consciousness and elicits only annoyance and boredom. After all, we have to presume that if Aksakov had strictly followed the family events of the decades preceding his birth, and moreover the events of his childhood, he could have filled dozens of volumes. He's incorrect in saying that free-ranging creativity is foreign to him. His creative work consisted of the skillful and masterly selection of material; a feeling for proportion; the conceptualization of the material, its arrangement and poetic expression; and also the typification of facts, and the imparting to them of a generalized moment. Facts about his everyday life had to be elevated to the level of art and, more than this, to the level of poetry. Thus far, no one had contested the poetic quality of Aksakov's prose.

In answering one of his critics, Aksakov himself stated: "Similarly, I can't agree with the notion that I 'limited the stories about the activities of Kurolyosov to no purpose,' and that I 'touch upon his actions only in more general descriptions.' Whether these descriptions are good or not is another matter, but I remain convinced that enough details about Kurolyosov are provided, and that if there were more of them, *it would violate the artistry of the impressions*" [my emphasis—V.S.].

And thus, the literary significance of Aksakov, who was valued for his merits by his contemporaries and beloved by successive generations right up to the present, is not open to doubt. I still need to say a few words about the popular and national significance of Aksakov's literary legacy.

No one would argue with the fact that in the youth of each new generation it is necessary to cultivate patriotic feelings and what is known as "a feeling for one's homeland." But this "feeling for one's homeland" is very complex, for it includes a feeling for the history of one's own country, a feeling for the future, an evaluation of the present, and, in addition—but not last in priority—a feeling for nature in one's homeland. For example, a novel is about war but bears the title *White Birch*.[15]

And on the whole it's like twice two, that the general and profound love for one's country includes as a component part a love for the land, for its meadows, lakes, and forests. If contemporary poetry doesn't contain enough examples for us, we can turn to unfading classical examples:

> I love my motherland, but with a strange love!
> My intellect cannot conquer it.
> Neither glory bought with blood,
> Nor peace full of proud confidence,
> Nor the sacred legends of dark antiquity
> Can change my cherished dreams.
> But I love—I know not why—
> The cold silence of her steppes,
> The boundless rustle of her forests,
> Her rivers overflowing like the seas;
> I love to race by carriage on a country road
> And, with the slow, piercing gaze of night's shadows,
> To meet on both sides, while longing for shelter,
> The flickering lights of doleful villages;
> I love the smoke of burning hay,
> A string of carts that slumbers in the steppe,
> And on a hill in a yellow cornfield
> A pair of birches shining white.[16]

If we were to try to examine carefully our own feeling for our homeland, we would find that this feeling in us is not spontaneous, but rather is organized and cultured, since it has been nurtured not only by the spontaneous contemplation of nature as such, but also cultivated by the sum total of a country's art, by its entire cultural tradition. Love for our homeland has been cultivated in us by Pushkin, Lermontov, Tyutchev, Fet, Turgenev, A. K. Tolstoi, Nekrasov, Lev Tolstoi, Blok, Esenin, Levitan, Polenov, Savrasov, Nesterov, Kuindzhi, Shishkin. . . . What, then, should we do—in the above ranks (and they are numerous) should we really assign the name of Sergei Timofeevich Aksakov to last place? Even today his work continues to cultivate patriotism and love for nature in us; he also helps us to develop a love for our homeland and for the country we love—we would lay down our lives for our country and not give it up, we would spill our blood for it and not betray it. That's how it was, say, three decades ago.

Now let me narrow the scope of our discussion and say a few words concerning my own feelings about Aksakov.

First, I think of the sheer pleasure he gave me, a reader, by means of the purity and originality of his language, its expressiveness, poetic quality, and the spiritual health that I described earlier.

Second, like all village boys, I developed from my childhood years an association with fishing poles and various kinds of gudgeons and roaches found in our Vorshcha River, but I discovered the real passion of a fisherman, which also means the real pleasure derived from this form of hunting, in Aksakov, in his *Notes about Fishing*. You see, it's impossible for me to imagine today how many dawns, how many mornings bestrewn with large and heavy dewdrops (like peas) I spent on the river. One has to pay for them with at least some kind of gratitude.

Third, Aksakov started to write a book that remained unwritten. Several copies of the first page have survived, entitled *Notes and Observations of a Mushroom Hunter*. If we may judge by the beginning of the book, it would have been of the same type as his fishing and hunting books, and possibly even better than they. But he wasn't able to write it. And thus, precisely in continuation of these pages, I wrote, in my time, a little book about mushrooms called *The Third Hunt*.[17]

And fourth . . . and fourth, I learned a great deal from him. If I chose epigraphs from Aksakov for both my *A Drop of Dew* and the aforementioned *The Third Hunt*, it indicates not only that I liked those sentences but, to a greater degree, that both Aksakov's method of apprehending reality artistically and his manner of expressing his thoughts and feelings had become important to me.

This is why, when the little voice of the employee Tanya of *Literaturnaya gazeta* suggested that I travel to Buguruslan, and I somewhat rudely asked,

"What haven't I seen in this Buguruslan?" and Tanya answered that I hadn't seen what condition Aksakov's former estate was in—this is why, after this sentence of Tanya's, I fell expressively silent, which gave her the opportunity to ask me, without the slightest interrogative intonation, "Then I'll send you the letter. . . . "

The letter turned out to be not the original, which had been sent from Orenburg Province to the writer, Nadezhda Vasilyievna Chertova. But since she was unable to address this matter adequately, due to the state of her health and her advancing years, she wrote a letter to *Literaturnaya gazeta* that was based on the letter from Orenburg. This second letter I now hold in my hands. Here it is, word for word:

ON THE CREATION OF A MEMORIAL COMPLEX
IN THE HAMLET OF AKSAKOVO OF ORENBURG PROVINCE

The hamlet of Aksakovo, where the well-known Russian writer Sergei Timofeevich Aksakov, spent his childhood, is located in Orenburg Province, on the bank of the Great Buguruslan River. Aksakov's *Orenburg Stories*, and especially *The Childhood Years of Bagrov the Grandson*, live on in the memory of grateful generations of Russian and Soviet readers.

Even today, thousands of "self-motivated" excursionists, mainly schoolchildren, walk or drive to the hamlet of Aksakovo—not only from the schools, cities, and villages of Orenburg Province, but also from neighboring provinces. And what do they see in the location of what S. T. Aksakov described so uniquely, his "dear Bagrov"?

Only thirty years ago the Grachev birch grove was *cut down* [Chertova's emphasis—V.N.]. The Aksakovs' house was demolished in 1960—a small, two-story school building was put up in its place. In 1966 the water mill on the bank of the Great Buguruslan River burned down and has not been restored. The park, which is not enclosed and not protected, remains neglected and is deteriorating into a state of ruin.

Only if one stands in a certain place beside the school building is it possible to reconstruct the view that at one time the eyes of the young Aksakov had seen, and recall the lines:

"I liked very much to look out the window that opened up onto the Buguruslan: from it I could see the expanses of forests growing out of the riverbed of the Buguruslan, where they merged with those of the Karmalka Stream, and between them the steep, bare peak of Mount Chelyaevsk."

As is well known, the writer considered it absolutely necessary "to convey to others [his] impressions with the accuracy and clarity of an eyewitness in such a way that readers would have the same perception of the events being described as I had of them."

The natural world still flourishes in the hamlet of Aksakovo. But it is imperative for us to finally complete the picture and restore what was lost to the precise form that inspired the writer to create his wonderful book.

This matter—of the establishment of a memorial complex in Aksakovo—has a long history, and to this day it hasn't gotten off the ground.

Back at the beginning of the 1950s, the people of the province raised the question of creating a museum-estate in Aksakovo, but at that time the matter did not proceed beyond the initial discussions. Finally, in May of 1971, the Executive Committee of Orenburg adopted a resolution for the creation of a memorial complex in Aksakovo. About four years have passed since that time, but the matter hasn't really gathered any momentum. To be sure, the Society for the Preservation of Historical and Cultural Monuments[18] paid for the renovation of one of the school buildings for a museum devoted to Aksakov, but a boarding school appropriated this building. The official designated by the Ministry of Culture (to maintain the museum) is occupied at the museum of the city of Buguruslan. Two years ago the students of the Aksakovo school planted six hundred pine trees in the park, and the saplings took root; but the park was not enclosed and cattle ate the entire plantation. Documentation was compiled for the cleaning of the pond, in light of which the collective farm Rodina requested that the need to provide water for four thousand large, horned cattle, and also the possible organization of a profitable fish industry, be taken into account. The cost of all this work was assessed as the sum of up to one million rubles. Naturally, such a large amount of money did not materialize, and the collective farm itself flatly refused to participate in any way in the project, citing the poor state of its own economy. For all these reasons the pond, too, was not set in order.

Approximately in 1969 an enthusiast of this undertaking—Honored Teacher of the Republic N. G. Khlebnikov—sent an article about Aksakov (concerning the creation of a museum) to *Literaturnaya Rossiya* and received an encouraging reply from the now deceased L. N. Fomenko. However, the editors themselves refused to publish it.

Some notices about the hamlet of Aksakovo appeared as a flash in the darkness: in *Pravda*, "Pod senyu parka" (Under the canopy of the park) by V. Shalgunov (July 27, 1974), and another one in *Sovetskaya kultura* (January 4, 1975).[19] However, most of all they resembled lyrical meditations and did not contain an urgent, businesslike formulation of the question. These publications did not produce any results.

Last year a commission that visited Aksakovo recommended that the Provincial Bureau of Excursions not send any tourists there until the territory is put in order in accordance with the resolution of the Executive Committee of Orenburg Province. At this point the matter finally came to a standstill, and meanwhile the tourists keep coming to Aksakovo all the same.

It is obvious that our central press must become involved in this matter, and most desirable for this role is *Literaturnaya gazeta*.

N. Chertova

(Compiled from materials received from Honored Teacher of the Republic Nikolai Gennadievich Khlebnikov.)

I read the letter but so far haven't answered either yes or no to it. I'll think about it. I'll consider it in light of my Moscow circumstances. And my head tells me almost resolutely that now, in late autumn, during the transition from autumn to winter, is hardly the time to drag myself to a Buguruslan unknown to me; but my heart no less resolutely decides for me: I have to go. "I left my grandfather, I left my grandmother. . . ."

"Listen . . ." (after lifting the receiver and dialing the number), "did you at least find out how long it takes to get there, on which train?"

"Yes, I did. You take the Karagandinsk train. From Kazan Station. You travel for exactly twenty-four hours and get off right in Buguruslan. From there you travel only about thirty kilometers—and you've reached the village of Aksakovo."

"Can I first fly to a big city instead?"

"I can tell you. From Orenburg, the provincial center, it's five hundred kilometers; from Ufa it's also . . . about three hundred; and if you fly to Kuibyshev . . . you still have to ride the train for several hours. But why do you have to travel that way? It's complicated to get from the airport to the train station. Wouldn't it be easier to get on in Moscow and get off in Buguruslan?"

"Well, all right. I'll think about it. I'll consider it in light of my Moscow circumstances."

In reality the consideration proceeded along two opposing paths. On the one hand, it really was the case that Moscow everyday circumstances almost always militated against any extra trip and in general against any extra workload, since one is constantly in a whirlwind of time running out, and on every day of the week something is scheduled that seems impossible to postpone or cancel. On the other hand, the consideration came from the stirrings in my soul, from something that responded from its place in my soul the instant the word "Aksakov" was uttered over the telephone—such a short word, one so well known to each person who in his or her life has read even just an excerpt from Aksakov's work. How could I not go—to Aksakov?

✦ ✦ ✦

HISTORICAL INFORMATION ABOUT THE HAMLET OF AKSAKOVO OF THE
BUGURUSLAN REGION AND ABOUT THE RESIDENCE IN IT OF
THE WRITER SERGEI TIMOFEEVICH AKSAKOV

The hamlet of Aksakovo, also called the hamlet of Znamenskoe because of
its church of the same name that was built in honor of the Feast of the Sign,[20]
was founded in the 1760s by Sergei Timofeevich's grandfather, Stepan
Mikhailovich Aksakov, a retired quartermaster and landowner from Simbirsk.
The little village that he founded was named after him: "Aksakovo." The land
was purchased from Nikolai Gryazev, a member of the Life Guards of the Pre-
obrazhensk Regiment and a bombardier (in *A Family Chronicle* S. T. Aksakov
writes that the land was bought from the [female—V.N.] landowner Gryaze-
va). In the 1780s the village of Aksakovo consisted of only about ten home-
steads of serfs.

After the death of Stepan Mikhailovich Aksakov, the village of Aksakovo
was passed down to his only son, Timofei Stepanovich, father of the future
writer Sergei Timofeevich.

The marriage between Timofei Stepanovich and Maria Nikolaevna Zubova
on September 20 (October 1, O.S.), 1791, in the city of Ufa produced the
child Sergei Timofeevich. He spent the years of his childhood and adolescence in what is now the hamlet of Aksakovo. These years and the natural
world all around him are superbly described by S. T. Aksakov in his works *The
Childhood Years of Bagrov the Grandson* and *A Family Chronicle*.

After studying in Kazan (from 1802 he was at the gymnasium, and then the
university), S. T. Aksakov resided in 1807 in both St. Petersburg and Moscow.
He married in 1816, and a year later settled in the hamlet of Aksakovo with his
family, where they remained until 1820. His first son, Konstantin Sergeevich—later the famous Slavophile—was born here, in Aksakovo.

In 1820 Sergei Timofeevich and his family moved to Moscow for a short
time, and in August of 1821 they once again returned to the territory of Orenburg Province and settled in the hamlet of Nadezhdino of Belebeevsk District,
which was described in *A Family Chronicle* under the name "Parashino." This
hamlet was part of his father's estate and was given to Sergei Timofeevich as an
inheritance in 1821. He left this region in the fall of 1826 to go to Moscow.

I don't have any materials attesting to whether or not Sergei Timofeevich
spent any time in Aksakovo after 1820, but evidently he and his family traveled to Orenburg Province each summer, and probably spent some time in
Aksakovo as well.

In 1849 S. T. Aksakov purchased the estate of Abramtsevo near Moscow, and from that time until his death he lived there.

S. T. Aksakov died on the night of April 30, 1859. I also don't have any materials concerning how the estate of Timofei Stepanovich Aksakov was divided after his death (he died in 1837) or which member of the Aksakov family inherited the hamlet of Aksakovo. But evidently Aksakovo passed into the hands of Arkady Timofeevich, brother of Sergei Timofeevich, since it had belonged to him during the era preceding the "Great Reforms."[21]

I don't have any evidence documenting the subsequent fate of the Aksakovs' estate in the hamlet of Aksakovo. During a trip to Aksakovo in the summer of 1958 I was informed by the old-timers there that the last owners of the estate in Aksakovo were Sergei Arkadievich Aksakov and his sons. In the old-timers' words, S. A. Aksakov had worked as a zemstvo chief[22] in prerevolutionary Russia.

During this trip to Aksakovo in 1958 I found the two-story wooden house still intact. It was in good condition. The house's upper story contained the apartments of workers of the Aksakovo MTS.[23] Several stone buildings stood near the house. The estate included a park. In some places the traces of paths could still be seen. A lane of linden trees was well preserved. Six old pine trees were still standing. Enormous white willow trees were growing on the bank of the Buguruslan.

The ancient pond was in a polluted state. There were no longer any fruit trees. Workers' vegetable gardens could be seen in the clearings of the park.

On the spot where the church had stood (it had been built by Sergei Timofeevich's father at the end of the eighteenth and beginning of the nineteenth centuries) lay a pile of road-metal and garbage, while next to this pile three gravestones lay in disorder.

One gravestone of black granite was shaped like a square and was about seventy centimeters high. It lay on its side. I could read the inscription carved on it: "Arkady Timofeevich Aksakov, born January 15, 1803, died October 15, 1862."

The second gravestone, cut from pink granite, was over a meter in length. I wasn't able to read the inscription on it.

The third gravestone was cut from gray granite and also more than a meter in length. The inscription on it had been damaged intentionally, but I could still read several words and letters:

Maria Nikolaevna Aksakova,
née ——————— ova
Born ——————— the 7th day of January
Died ——————— .... day ....

These gravestones indisputably marked the graves of Arkady Timofeevich Aksakov, the writer's brother, and Maria Nikolaevna Zubova, Sergei Timofeevich's mother. The third, in all likelihood, marked the grave of the writer's father, Timofei Sergeevich.

Research Associate,
Provincial Museum   A. S. Popov
November 24, 1968

◆ ◆ ◆

ALL-RUSSIAN SOCIETY FOR THE PROTECTION OF HISTORICAL
AND CULTURAL MONUMENTS DIVISION OF ORENBURG
PROVINCE CITY OF ORENBURG

May 20, 1971

To the First Secretary of the Buguruslan
City Soviet of the CPSS, T.[24] Karpets

Respected Vasily Andreevich,

At the beginning of February of this year I was in Moscow for the plenum of the Board of the Society for the Preservation of Historical and Cultural Monuments.

In the presentation of T. Ivanov (acting chairman of the board) our province was subjected to rather harsh criticism for the destruction of Aksakov's estate. This criticism was supported by T. Kochemasov, Vyacheslav Ivanovich—Acting Chairman of the Soviet of Ministers of the R.S.F.S.R.

In their presentations they proposed the task of preserving and restoring what still remains intact of the estate, and of putting the estate's lands in order.

You evidently know that a representative from the Ministry of Culture of the R.S.F.S.R. came to Buguruslan and Aksakovo, and photographed the generally unattractive condition of the estate's remains, the oppressive state of the neglected garden, and the disgracefully violated gravestones of S. T. Aksakov's relatives.

These photographs serve as an irrefutable reproach to all of us; they can be found at the Ministry of Culture and the board of the Society, whose chairman is T. Kochemasov.

Apparently, considerable sums of money were released several times for the partial restoration of the estate, which has been acknowledged as a monument of

50

national significance; but instead of the restoration of what had remained, the further destruction of the estate proceeded.

Now the task has been proposed to create a memorial monument to a great writer of the Russian land, a poet of the wonderful natural setting of the Orenburg region. This undertaking would include the restoration of the pond; fortification of the banks of the Buguruslan River; prevention of the ruin of one part of the park, namely, the lane of linden trees; enclosure of the park; replacement of the gravestones; erection of a monument to the writer; and establishment of a museum in the building that houses the boarding school (incidentally, this building belongs to the state, not to the collective farm).

The central administration issued twenty thousand rubles to us for these purposes. On May 7th I told Alexander Vlasovich all of this. He approved all the goals listed in connection with this project. The proposal to decide on the partial restoration of Aksakov's estate has been submitted to the Provincial Executive Committee for its consideration, and evidently it will be accepted next week.

In connection with the forthcoming large-scale efforts to carry out this resolution of the Provincial Executive Committee, I have an earnest request to make of you and the members of the bureau: should we discuss this matter at the meeting of the bureau (which would be the most expedient alternative) or at the meeting of the executive committee? We need to request that the local organizations commence work, so that in the near future even a minimum of the work to put the estate in order will be accomplished.

I consider the opportunity to restore what remains of the destroyed estate to be a matter of honor. After all, it will constitute a cultural hearth important to the republic. I dare say that in discussing this matter you won't omit the task of the economic fortification and aesthetic improvement of the collective farm Rodina, which evidently will have to receive preferential help, since tourists (and with each year there will be more of them) will want to take a look as well at the collective farm, its cultural and industrial buildings, and the everyday life of the collective farm workers.

I'd like very much to come to Buguruslan and present some lectures and recollections about Lenin, but at the moment my health prevents me from doing so. But I haven't lost hope that I'll be able to visit your city again.

With a Communist's greeting,
A. Bocharov,
Member of the CPSS since 1920 and Candidate of the Historical Sciences

Whereas the letter of the honored teacher of the republic was presented to me while I was still in Moscow, the two latter documents I obtained already on site, in Buguruslan, in the form of typed copies containing round holes for small cord by means of which the papers are filed into folders. Moreover, upon removal of the copies, the small holes were definitely torn by someone. I also acquired other documents: acts, resolutions, decrees, memoranda, inquiries, and responses to inquiries. The following picture emerged.

In the 1920s the Aksakovs' house initially housed an MTS. An office and habitable rooms were also located there. The vegetable gardens in the clearings of the park, which figured in the historical account of A. S. Popov, were planted and cultivated specifically during this period.

Later on, the MTSs, as is well known, were abolished throughout the country. The house was left without an owner, even though it stood on the territory of the collective farm Rodina. The Provincial Executive Committee allocated funds several times for the repair of the house and even allocated materials as well, in particular, roofing iron. But the collective farm (i.e., its chairman, Ivan Alexandrovich Markov) was too busy to repair the house; although the house did not belong to it, besides the collective farm there was no one to repair the house. The materials and money were used for other, perhaps even important, purposes.

Finally, the collective farm began to petition for the house and entire estate to be placed completely under its jurisdiction. And when this occurred, in 1960 (that is, very, very recently), the collective farm, having become the owner of Aksakov's house, demolished this house and on its foundation built a school, also two stories high, approximately of the same overall dimensions. The fact that the house didn't have a roof doomed it to rapid destruction. In describing the house's condition, the chairman of the collective farm, I. A. Markov, himself told me the following: "It was in a disastrous state. The attic and upper story were firmly packed with snow, and then the snow began to melt. . . . "

The idea of the school was very cleverly conceived, since who would later object that a house had been replaced by a school? Indeed. At this point I recall the brief review in *Yuzhnyi Ural*, and shall quote the second half of it, as I have promised. To provide continuity, let's begin with a sentence I've already cited. Thus:

> Not in a certain kingdom, but in the homeland of the housekeeper Pelageya, storytelling is carried out these days by compatriots of Aksakov—the collective farm workers of the agricultural artel Rodina. A session [i.e., a storytelling session— V.S.] lasted for several days. In place of the ancient, disintegrating house in which the childhood years of the prominent author of *A Family Chronicle* had been spent, a two-story stone school building was erected, financed by the collective farm.

Well, we'll leave it to the conscience of the author of the review to decide whether or not the construction of a typical school during the second half of the twentieth century in our enlightened state is like something from out of a fairy tale. But the fact that he put in the word "disintegrating" with respect to the house cannot be permitted to remain on anyone's conscience. Both the collective farm workers of the hamlet of Aksakovo and the head of the school's academic division, Andrei Pavlovich Tovpeko, told me the same thing: the house was amazingly solid. The logs in it had been brought long ago from the Buzuluksk pine forest: they had been carefully selected and were larger around than a man could embrace with both arms. And besides, the forest had been maintained properly in the ancient way, while the rows of logs had been installed on top of special tenons. So when they demolished the house, it was log by log, and each row of logs had to be torn away by a tractor. The roof, of course, was bad, and because of it the ceilings were bad as well. But now we know that the roof had not been repaired, while money and even roofing iron had been allotted several times for its repair.

Incidentally, I have always been struck by our love for building something "in place of" rather than "beside." In order to build a school they were obliged to demolish a house. Why? For what purpose? Why not locate the school next to the house? The aforementioned Andrei Pavlovich Tovpeko told me that it had been unwise to erect the school on the old foundation of Aksakov's house, since it restricted its overall dimensions; the internal layout of the school's rooms—that is, classrooms—is now cramped.

I mentioned the numerous documents (photocopies, of course, with small holes in them) that I had acquired.[25]

Perhaps I didn't have to take them out for all the world to see, but in any case, in order to illustrate and confirm the "fairy-tale-like" actions that occurred there, I need to include several of these documents in this essay.

RESOLUTION OF THE PROVINCIAL EXECUTIVE COMMITTEE OF
SEPTEMBER 4, 1953 ON THE STATE OF THE PRESERVATION
OF HISTORICAL MONUMENTS

The extremely valuable cultural monuments of the former country estate of the writer S. T. Aksakov, which is included in the balance sheets of the Mordovo-Bokminsk MTS, are being destroyed. On September 25, 1953, the Mordovo-Bokminsk region decided to register all of the historical *monuments and ensure their protection.*

1. During the years 1953–54, repairs of the historical monuments should be completed at the expense of the local budget for the improvement of the region.

The director of the Mordovo-Bokminsk MTS, T. Lyubakov, should be required from 1953 to 1954 to complete the repair and restoration of the historical monument of the former country estate of S. T. Aksakov.

2. The head of the Provincial Administration of Agriculture and State Purchases, T. Dushenkov, should allocate in a timely manner MTS funds for the renovation work, and should maintain daily control over the restoration of this building.

3. The Provincial Department of Culture should be given control over the administration of the accounts for the protection, restoration, and repair of the historical monument.

A. Zhukov
Chairman of the Provincial Executive Committee

B. Beidyukov
Secretary of the Provincial Executive Committee

A fine, useful resolution. Perhaps the director of the MTS, T. Lyubakov, would even have adhered to it and have carried out the repairs on the building, especially since it was ordered that the head of the Provincial Administration of Agriculture, T. Dushenkov, "allocate in a timely manner MTS funds for the renovation work." But, as we know, the MTSs were abolished. The MTSs no longer exist, and there's no one we can ask about these things.

A new concern emerged: what to do with the unfortunate building that was left without an owner. As we know, there is an owner, of course. The owner, as we know, is—the people. But formally, on whose balance sheets is the house listed?

A new and likewise sensible resolution followed, passed by the Provincial Executive Committee on August 11, now in the year 1959, one year, as we now know, before the final destruction of the house.

RESOLUTION OF THE EXECUTIVE COMMITTEE OF THE ORENBURG
PROVINCIAL SOVIET ON THE USE OF THE FORMER ESTATE
OF THE WRITER S. T. AKSAKOV

The Executive Committee resolves:

1. To establish a dormitory in the building of S. T. Aksakov's former estate, and a boarding school in the auxiliary buildings.

2. To require the Provincial Plan to procure a maximum sum of 15,000 rubles for the compilation of technical documentation, and for the repair and reequipping of the building of the former estate of the writer S. T. Aksakov, and the

attachment to it of a new school building with 320 seats.

3.   To require the Provincial Project to compile technical documentation for the construction of the new school building by September 15, 1959. To carry out the repair and reequipping of the building of the writer's former estate in such a way as to preserve the architecture of the main residential house.

4.   To require the Provincial Department of Culture and the Buguruslan Regional Executive Committee to establish a museum of the writer S. T. Aksakov, and to this end to select and equip a room in the main building of the former estate.

5.   To require the Executive Committee of the Buguruslan Regional Soviet to take the necessary steps for the improvement of the grounds of the estate and garden, and to enlist the services of the community and the school in order to reach this goal.

A. Zhukov
Chairman of the Provincial Executive Committee

It would seem that nothing else was needed. They planned to preserve the building, carry out repairs in such a way as to preserve the architecture of the main residential house, and take steps to improve the grounds of the estate and garden. After all, what remained was only to execute this wonderful resolution, and then the gratitude of our descendants, not to mention that of our contemporaries, would be assured. But how on earth did it happen that exactly one year after this resolution Aksakov's house was razed?

There are two sides to the matter. Of course, without the consent of the province, without an OK, Ivan Alexandrovich Markov wouldn't have dared to tear down the house. On the other hand, without the animated desire and petition of Ivan Alexandrovich it wouldn't have occurred to the Provincial Executive Committee to tear down the house. As if the Provincial Executive Committee didn't have any other matters and concerns to attend to. And the house continued to stand, especially since the resolution for its preservation and repair had been passed. But if a burning request, buttressed by convincing arguments, comes from below, the Provincial Executive Committee can comply with the convincing request of the chairman of a collective farm. Evidently, the initiative to raze the house came from the chairman of the collective farm, Ivan Alexandrovich Markov. And the commission could be convinced that the house was dilapidated, that the roof and ceiling leaked, and that little kids crawled into it to play and could be crushed. It was precisely this reason that Ivan Alexandrovich cited to me personally as the main one. The house had become dangerous. Little kids played there. An accident could happen.

"And what if they fixed it?"

55

"That would be more complicated."

"Than demolishing it?"

"Why do you keep harping on this house? You know, we built a school in its place!"

The demolition of the house was preceded by a whole series of documents; I happen to have several of them in my hands.

LETTER TO THE CHAIRMAN OF THE PROVINCIAL EXECUTIVE COMMITTEE
FROM THE COLLECTIVE FARM RODINA

The Buguruslan Regional Executive Committee, based on the resolution of the general meeting of the collective farm workers on April 7, 1961, and of the party committee of the collective farm Rodina on April 5th, requests that the Aksakovo RTS[26] transfer from its balance sheets to those of the collective farm the estate of S. T. Aksakov, along with its plot of land, its household structures, and the writer's house; and also that the construction of an eleven-year school on the spot of the dilapidated house of S. T. Aksakov be permitted, using the funds of the collective farm.

This little document produced a chain of other documents, only three of which I have in front of me. First, after receiving the request of the collective farm that was corroborated by the region,[27] the Provincial Executive Committee made an inquiry at its own Department of Culture, to which it received the reply: "To your number [such-and-such—V.S.] of April 24, 1961, the Department of Culture informs you that the country estate of S. T. Aksakov is not listed in the register as a historical monument. Head of the Provincial Department of Cultural Affairs, V. Biryukov."

Well, if it's not listed, then what's there to talk about? But even so, the Regional Committee and Provincial Executive Committee, in all fairness, created and sent off a special commission so that it could confirm everything on site, sort things out, and make its own recommendations. The following recommendations by the commission were issued on August 1, 1961:

MEMORANDUM TO THE SECRETARY OF THE PROVINCIAL COMMITTEE OF
THE CPSS, T. SHURYGIN, V. N. AND TO THE CHAIRMAN OF THE REGIONAL
SOVIET, MOLCHANINOV

The commission consisting of the head of the Provincial Department of Public Education, Tkacheva; the manager of the Provincial Planning Office, T. Ivanov; and the chief engineer of the UKS[28] of the Provincial Executive Committee, T.

56

Trachtenberg, submits:

1.   That the organization of a boarding school on the territory of the country estate of the writer S. T. Aksakov is considered inadvisable, since the residential house and all of the remaining structures have fallen into a state of disrepair, and for the renovation of the residential house alone not less than 60,000 rubles will be needed.

2.   That after its restoration the residential house cannot be utilized completely, due to the fact that its second story is very low, being about one and one half meters high (!).   The other structures, in particular the stable, cannot be adapted for residential purposes because of the absence of minimal sanitary facilities.

3.   That the soundest course of action would be to transfer Aksakov's estate to the agricultural artel Rodina, and on its territory to construct (according to the standard plan) a school for general education.

The collective farm intends to proceed with the construction of the school and has allocated the necessary resources.   For a more rapid realization of the plan to construct the school, it would be appropriate to provide help during the construction process by allocating the necessary building materials.

The Executive Committee of the Buguruslan Regional Soviet, on August 14, 1961, requests that the Provincial Executive Committee support the petition of the administration of the collective farm Rodina.

Based on this memorandum, the Provincial Executive Committee issued its resolution soon thereafter.

RESOLUTION OF THE PROVINCIAL EXECUTIVE COMMITTEE ON THE CULTURAL MONUMENT OF THE WRITER SERGEI TIMOFEEVICH AKSAKOV

For the purpose of preserving the cultural monument of local significance, the former country estate of the writer Aksakov, and taking into account the petition of the administration of the collective farm Rodina, the Executive Committee of the Provincial Soviet resolves:

1.   To approve the petition of the collective farm Rodina of the Buguruslan region for the transfer to the collective farm of the land adjoining the estate, park, and existing buildings of the former country estate of the writer S. T. Aksakov, in order to put in order the territory of the former estate and to construct on it a standard school for general education.

2.   That the Provincial Department of Culture conclude with the collective farm *an agreement for the estate's preservation.*

3. To require that the Provincial Department of Culture (T. Biryukov) install on the building of the new school a memorial plaque for the perpetuation of the memory of the Russian writer, S. T. Aksakov.

N. Molchaninov
Chairman of the Executive Committee

A. Krasnov
Secretary of the Executive Committee

Let us note that in this document Aksakov's house is not even mentioned, as had been the case in the other documents from this same period. Here the statements are circuitous and mild concerning "the transfer to the collective farm of the land adjoining the estate, park, and *existing* buildings of the former country estate of the writer S. T. Aksakov" [emphasis mine—V.S.]. Evidently, news had reached the provincial administration that the house had already been torn down. All that remained was to install a memorial plaque on the building of the new school.

What happened to the house is relatively clear. But there still remained the park, pond, and mill. The park was a complicated matter. If a park is not cared for every day, from one year to the next, it becomes wild and, in practical terms, comes to ruin. The old trees fall down or are chopped down for firewood, while the shrubbery becomes overgrown and entangled, transforming what had been a park into something resembling an enormous, formless wisp of bast, from which only here and there the centuries-old trees can accidentally protrude. This is precisely what happened to the park at Aksakovo. I need only add that the main portion of old trees was cut down during the war years, when there were almost no peasant men left in the villages, at least no strong and healthy men, and so people obtained firewood where it was closest and most convenient.[29]

Regarding the mill and pond, we need to recall how they came to be. Fortunately, there is a marvelous and detailed description of this event in the very first pages of *A Family Chronicle*. Since thus far I haven't taken advantage of Aksakov's texts proper, I'll cite the description of how the mill was established. It is also of interest in light of the fact that, as it turns out, our ancestors used the same method of damming rivers as we do today when we construct dams across the Enisei or the Angara. To be sure, the proportions are not the same, nor are the materials, technology, and purposes. We no longer use brushwood or manure or straw, but rather reinforced concrete blocks; we don't use carts, but rather dump trucks; and we don't use one hundred able men, but rather a huge army of construction workers. But note that the principle remains the same:

> ... After having chosen a spot earlier, where the water wasn't deep, the river bottom was solid, and the banks high and also solid, they brought to it from both sides of the river a dam of brushwood and earth, like two hands ready to catch something. For greater strength they interwove into the dam flexible willow branches in the form of a wattle fence. What remained was to hold back the rapid and powerful water, and force it to fill the reservoir intended for it. On one side, where the bank seemed somewhat lower, a mill granary had been erected earlier on two flour-grinding tables with a millstone. All of the ropes and pulleys were ready and even oiled; on enormous waterwheels, through the wooden pipes of the reservoir the water was supposed to rush in, where, stopped in its own natural riverbed, it would fill a broad pond and become higher than the bottom of the reservoir. When everything was already ready and four long oak piles had been driven firmly into the hard, clay bottom of the Buguruslan, across the future enclosure that would release excess water from spring floods, grandfather arranged for work to be done for two days in exchange for food.[30] Our neighbors had been invited, and they brought their horses, wagons, spades, pitchforks, and axes. On the first day, huge piles of brushwood from bushes and a small forest that had been cut down, shocks of hay, manure, and fresh sod were heaped up on both sides of the Buguruslan, thus far free and inviolably aiming its waters. On the second day at sunrise, about one hundred people gathered to stop some water, that is, to dam up a river. Every face showed traces of concern and solemnity; everyone was preparing for something; the entire village virtually didn't sleep on this night. All together, at the exact same instant, with a loud crash, they moved into the river from both banks piles of brushwood that first had been tied into small piles; much was carried away by the water's swift current, but a great deal that was caught by the piles settled across the river bottom; the bundled shocks of hay and stones sank to the bottom as well, and they were followed by the manure and earth; once again the people put down a layer of brushwood, and again hay and manure, and on top of all this some thick layers of sod. When all of this, somehow submerged, had risen higher than the water's surface, about twenty peasants, all stalwart and agile, scrambled onto the top of the dam and began to trample and tread it down with their feet. All of this took place with such speed, such general zeal and uninterrupted shouting, that anyone driving or walking by would have been frightened upon hearing it, if he or she hadn't known the reason for the commotion. But there was no one who could be afraid: only the wild steppes and dark forests for many miles around resounded with the furious cries of one hundred workers, to which were added a multitude of women's

voices and even more voices of children, since everyone took part in such an important event, everyone bustled about, ran, and cried out. It took them a long time to bring the stubborn river under control: for a long time it tore and carried away the brushwood, hay, manure, and sod. But finally the people gained the upper hand: the water couldn't break through anymore, then stopped as if lost in thought, whirled around, went back, filled the banks of its channel, inundated and overflowed them, and began to flow out onto the meadows. By evening a pond had already been formed, or, to put it a better way, a lake had risen to the surface, without banks, and without the verdure, grasses, and bushes that always grow on them; here and there the tops of submerged, doomed trees stuck out. On the second day the millstone began to rotate, and the mill began to grind—and it mills and grinds to this day.

I don't know the year when the mill stopped revolving and milling, but the structure itself—the mill granary, reservoir, and wheels—all of it burned down in 1966, having outlived the Aksakov house by six years. The pond didn't burn down at this time, as one might surmise, but it hadn't been cleaned or washed out thoroughly perhaps since Aksakov's time; it is polluted, filled with silt, has grown shallow and overgrown with plants, no longer contains any fish, and has turned into a huge puddle.

I don't know why people always called and still call it a pond. It is more properly a "mill pool," a "storage pool," or "reservoir" that enhances and ennobles an area of steppes. And if it were to be cleaned up by removing all the silt to the fields of the collective farm Rodina, and properly and skillfully stocked with fish, if a minimum of effort were to be exerted toward the maintenance of purity and order in it, then it would even be of economic significance.

Well, you see, this means that on all points there is complete correspondence: the park has grown wild, the pond has been neglected, the house was razed, and the mill burned down. The perfect time to undertake the protection and restoration of a so-called memorial complex.

✦ ✦ ✦

THE ORENBURG PROVINCIAL DIVISION OF THE ALL-RUSSIAN SOCIETY
FOR THE PRESERVATION OF MONUMENTS
CITY OF ORENBURG, 66 SOVETSKAYA ST., ROOM 68

The State Inspection for the Preservation of Historical and Cultural Monuments announces that the former country estate of S. T. Aksakov in the Buguruslan region of Orenburg Province has been included in the register of historical monuments under government protection.

Regarding this matter, we ask you to petition the Provincial Executive Committee to carry out urgent steps for the preservation of the memorial park, and likewise to set aside a location for the museum of S. T. Aksakov. According to correspondence that we have received, materials for the museum can be provided by the museum-estate "Abramtsevo."

Head of State Inspection
for the Preservation of Historical and
Cultural Monuments

(Makovetsky)

✦ ✦ ✦

TO THE DIRECTOR OF THE MEMORIAL MUSEUM
"ABRAMTSEVO," T. MANIN, V. F.

In May of 1971 the Executive Committee of the Orenburg Provincial Soviet of Workers' Deputies passed a resolution, "On the creation of a memorial complex for the writer, Sergei Timofeevich Aksakov, in the hamlet of Aksakovo of the Buguruslan region."

According to the Executive Committee's resolution, project-allowance organizations are required in order to develop a general plan for restoration work on the former country estate of S. T. Aksakov. The project assignments listed in the general plan for the restoration and repair work stipulate the following: restoration of the house on Aksakov's estate; renovation of the park; clearing of the existing plantations and planting of valuable species of trees; creation of arbors and walkways; layout of a small, sunken public garden; and restoration of the pond with its water mill, dam, and derivation canal.

Projections will require the following: photographs; drawings; sketches; descriptions of Aksakov's house, mill, pond, park, and arbors; and other materials. Our provincial division does not have these materials in its possession.

In order to help the planners achieve the most complete restoration of the memorial complex to its former state, I earnestly request that you indicate where and how the necessary materials for the country estate of the writer Aksakov can be located.

And if you have in your museum any photographs; sketches; drawings; descriptions of Aksakov's house, mill, pond, park, and arbors; and other materials, would you be so kind as to send copies of them to the address of the provincial division of the Society:[31] City of Orenburg, 66 Sovetskaya St., Room 68.

Chairman of the Presidium of the Provincial Division of the Society
(A. Bochagov)

✦ ✦ ✦

DESIGN ASSIGNMENT FOR THE INSTITUTE OF ORENBURG AGRICULTURAL PROJECTS

On the basis of the protocol of the meeting on August 17, 1970, of the Presidium of the Orenburg Provincial Division of the All-Russian Society for the Preservation of Historical and Cultural Monuments, and of the investigation of the monuments and memorial places of the Buguruslan region on November 12, 1968, we find it necessary to compile project-allowance documentation for the restoration of the country estate of Aksakov.

Compilation of the project-allowance documentation should provide for the following:

1. Enclosure of the park (an iron fence on supports of reinforced concrete).

2. Clearing of the existing plantations and planting of valuable species of trees.

3. Cleaning out and renovation of the pond, including fish-ponds for the breeding of fish.

4. Fortification of the banks of the Buguruslan River.

5. Preservation and repair of the five existing brick buildings.

6. Construction of a memorial complex that would include a hotel for tourists, dining hall, and a memorial room dedicated to Aksakov.

7. Placement of the gravestones from the graves of Aksakov's parents and the restoration of the inscriptions on the gravestones.

Compilation of the general plan of the hamlet of Aksakovo should provide for the preservation of the memorial park, and its inclusion in the rest area of the central country estate of the collective farm.

Payment for the compilation of the project-allowance documentation will be issued by the Orenburg Division of the All-Russian Society for the Preservation of Historical and Cultural Monuments.

A. Bochagov
Chairman of the Provincial Division of VOOPIK

I. Markov
Chairman of the Collective Farm Rodina

◆ ◆ ◆

RESOLUTION OF THE EXECUTIVE COMMITTEE OF THE ORENBURG PROVINCIAL SOVIET OF WORKERS' DEPUTIES

May 26, 1971

Concerning the Establishment of a Memorial Complex
of the Writer Sergei Timofeevich Aksakov in the
Hamlet of Aksakovo of the Buguruslan Region

October of 1971 marks the 180th anniversary of the birth of the Russian writer S. T. Aksakov, who for a long period of time lived and worked in the Orenburg territory; taking into account his enormous contributions to the development of culture and his popularity among Russian and foreign readers, for the purpose of the perpetuation of his memory, the Executive Committee of the Provincial Soviet R E S O L V E S:

1. To establish in the hamlet of Aksakovo, on the territory of the writer's former country estate, a memorial complex for S. T. Aksakov. To include in the memorial complex all of the buildings that belonged to S. T. Aksakov, the park, museum, and monument to the writer. To preserve the gravestones from the graves of the writer's parents and brother.

2. To require the head of the provincial planning office of the Provincial Administrative Committee of Economic Affairs, N. I. Belyaev, to include in the plan and projection for 1972: the development of the general plan for the restoration and repair work of the former country estate of S. T. Aksakov; the compilation for 1971 of project-allowance documentation for the renovation of the house as a museum to S. T. Aksakov; and the placement of a monument and gravestones of the parents and brother of S. T. Aksakov.

3. To require the director of the Institute of Orenburg Agricultural Projects, G. A. Reshyotnikov, in the compilation of the general plan of the construction of

the hamlet of Aksakovo (the collective farm Rodina) to take into account the obligation to preserve S. T. Aksakov's country estate with all its buildings and its park. And not later than July of this year to determine, along with the Provincial Division of the Society for the Preservation of Historical and Cultural Monuments, the borders of the writer's estate, as well as a protected zone.

Payment for the cost of the project-allowance documentation and renovation work on the house-museum, and the placement of a monument and the grave-stones of S. T. Aksakov's parents should be made at the expense of the Provincial Division of the Society for the Preservation of Historical and Cultural Monuments.

4. To require the Provincial Department of Repairs, Construction, and Restoration (T. Chekmaryov, S. S.) during the year 1971 to execute the major construction work on the creation of the memorial complex in the hamlet of Aksakovo. The Provincial Division of the Society for the Preservation of Historical and Cultural Monuments should conclude an agreement with the Provincial Department of Repairs, Construction, and Restoration to execute the restoration work and provide the Department with the financing.

5. To require the Provincial Division of the Society for the Preservation of Historical and Cultural Monuments (T. Bochagov, A. K.) by July 15, 1971, to conclude an agreement with the collective farm Rodina for the protection of the structures that have been transferred to the latter to be used for economic purposes.

6. To require the Buguruslan Regional Executive Committee (T. Proskurin, V. D.) to do the following:

(a) not later than July of this year to decide the matter of vacating one of the buildings, now occupied by a boarding school, for the purpose of establishing in it a museum devoted to the writer's life and work;

(b) to ensure the preservation of all the buildings remaining on the writer's country estate that were transferred to the collective farm Rodina.

(c) to improve the routes of entrance into the hamlet of Aksakovo.

7. To require the Provincial Department of Culture (T. Solovyov, A. V.) to petition the Ministry of Culture of the R.S.F.S.R for the opening of a branch of a museum of S. T. Aksakov.

8. To require the Provincial Soviet for Tourism (T. Pustovalov, M. F.) to develop an excursion itinerary for Aksakovo by 1972; to consider the establishment of a tourist base in the hamlet of Aksakovo; and, together with the Provincial Division of the Society for the Preservation of Historical and Cultural Monuments, to publish a guidebook of places in and around Aksakovo.

9. To require the Provincial Consumption Union (T. Serbin, G. P.) to decide the matter of the construction in 1972 of a dining hall with 25–30 seats in the

hamlet of Aksakovo, and to stipulate in the plan deliveries of 20–30 prefabricated houses to be sold to the inhabitants of the hamlet of Aksakovo.

10. To require the Provincial Administration of Forestry (T. Nechaev, N. A.) to carry out in 1971 the necessary renovation work in the park of the hamlet of Aksakovo.

11. To request that the Provincial Division of the Society for the Preservation of Nature (T. Vlasyuk, A. E.) take the park in Aksakov's country estate under its protection.

12. To charge the Orenburg Branch of the State Institute for the Planning of Water Management of the Central Volga[32] (T. Tafintsev, A. G.) to compile project-allowance documentation for the restoration work in 1971 of the pond in the park, as funded to the limit by the Provincial Department of Water Management.

13. To require the Provincial Department of Melioration and Water Management (T. Bomov, P.I.) to execute all of the restoration work of the pond in the park.

14. To ask the committee on matters of the press under the Soviet of Ministers of the R.S.F.S.R. about reissuing the works of S. T. Aksakov.

15. To request that the Regional Committee of the All-Union Leninist Young Communist League of the Soviet Union[33] (T. Zelepukhin, A. G.) provide a student force of construction workers for the period of the restoration work in the hamlet of Aksakovo.

16. To require the Provincial Department of Culture (T. Solovyov, A. V.) and the Provincial Division of the Society for the Preservation of Historical and Cultural Monuments (T. Bochagov, A. K.) to assume control over the completion of the work on the establishment of a memorial complex in the hamlet of Aksakovo and the equipping of the house-museum, and likewise to decide jointly the matter of providing an official employee (a museum worker) for the period of its repair and organization.

A. Balandin
Chairman of the Executive Committee of the
Provincial Soviet of Workers' Deputies

A. Karpunkov
Secretary of the Executive Committee of the
Provincial Soviet of Workers' Deputies

Z. Chaplygina
Attested: Head of the Protocol Section

This document was sent to: the Orenburg Agricultural Project, the Provincial Department of Repairs and Construction, the Provincial Soviet for Tourism, the Provincial Consumption Union, the Provincial Municipal Administration, the Provincial Department of Melioration and Water Management, the Provincial Department of Culture, the Society for the Preservation of Historical and Cultural Monuments, the Provincial Department for the Preservation of Nature, the Provincial Committee of the All-Union Leninist Young Communist League of the Soviet Union, the Provincial Administration of the Press, the Provincial Department of Matters of Construction and Architecture, T. Chernysheva, the Provincial Plan, the Provincial Department of Finance, the Provincial Committee of the CPSS, the Provincial Procurator T. Vlasyuk, the Buguruslan Regional Executive Committee, the collective farm Rodina of the Buguruslan region, T. Karpets of the Buguruslan City Committee of the CPSS, and the Orenburg Branch of the State Institute for the Planning of Water Management of the Central Volga.

After everything that has been stated, it's not difficult to imagine what I saw and found in Aksakovo.

In Buguruslan, that is, on the regional level, people treated me well and attentively, just like they would treat any guest from Moscow, and moreover one who also had with him a document from *Literaturnaya gazeta*. Incidentally, my impressions of Buguruslan are irrelevant here because they no longer relate to the subject of Aksakov, or more precisely, they do not constitute the subject of Aksakov in its pure form. For this reason, I'll say only that they provided me with a car for the trip to Aksakovo, and also some traveling companions: someone from the Regional Executive Committee, someone from the local newspaper, and another person—I just can't remember now which particular organization he represented. In a word, the GAZ[34] car of the latest model was packed to bursting, and we drove off.

On this day a session of the Regional Executive Committee was taking place, and the chairman of the collective farm Rodina, I. A. Markov, was expected to be present. And we were expected to wait for him in Aksakovo, for he had promised to arrive not later than two P.M., that is, by dinner. This meant that until two o'clock we could familiarize ourselves independently with the object of our visit. Incidentally, they thought that I was in Aksakovo for the first time. But I had stayed in Buguruslan for three whole days, waiting for them to give me a car. As if I could have sat calmly in my hotel for three days! In the meantime, on the second day, an "entrepreneur" drove me to Aksakovo for a five-ruble note, took me around the hamlet, waited while I walked around and asked some questions, and drove me back to Buguruslan.

But our current trip distinguished itself not only, so to speak, by its legality and officialness, but also by the fact that we had planned to drive to Aksakovo from the other side of the Buguruslan region, to traverse a big circle in order to come out onto the old Ufimsk Road, and along it virtually to repeat the oft-traveled route of Aksakov himself from Ufa to his native hamlet.

It turned out to be a marvelous day, as if we had placed an order: quiet, sunny—unusual in this region for the end of October. Around us two colors predominated: deep blue and gold. The clear sky was deep blue, while the hills that spread out under the sky, and also the sun—large and sharply outlined against the rich blue color—were golden. Of course, at times the hills were a reddish color, which was typical for this region, and at times among the autumn gold the rectangles of ploughed black earth appeared, clear and velvety-black. Of course, the forests on the hills and in the depressions between hills had already lost the greater part of their foliage and now were blackish, except for the oak groves, which as before were a coppery red, as if they had been cast and minted. But even the black, bare forest appeared golden under the clear autumn sun. There was also another kind of diversity: of fields and hamlets, roads, poles along both sides of the road, and oil towers here and there. But nevertheless, now when I try to recall the picturesque scenery of that day, I see two basic, predominating colors—deep blue and gold.

The road kept leading us through a sharply delineated landscape: from a hill into a deep ravine, obliquely along a hillside, and from out of a deep hollow onto a hill. At last, from a rounded ridge we saw below—in truth, as in the palm of a hand or on a tray—a large hamlet, in the overall view of which the straight rows of small, prefabricated houses with slate roofs, evidently built very recently, stood out. There were several dozen of them here, and I recall that I immediately made a note to myself, knowing the appropriate price of each house of this kind, that the collective farm Rodina was not poor at all, and that I needed to link what I had seen with the lines from the initial letter that, as the saying goes, had impelled me to make this official trip: "Documentation was compiled for the cleaning of the pond, in light of which the collective farm Rodina requested that the need to provide water for four thousand large, horned cattle, and also the possible organization of a profitable fish industry, be taken into account. The cost of all this work was expressed as the sum of up to one million rubles. Naturally, such a large amount of money did not materialize, and the collective farm itself flatly refused to participate in any way in the project, citing the poor state of its own economy."

But I must say initially that, upon first looking at Aksakovo from a high mountain, I sensed that something was missing there, that in some way this view was atypical. It goes without saying that thus far I had seen the hamlet from this high place only in pictures occasionally reproduced in Aksakov's books or in books written about him. My gaze had gotten used to the hamlet's appearance, but now when I looked something seemed to be missing from its usual appearance. It was just as if one were to view Moscow from a vantage point, and suddenly there was no Kremlin; in place of the Kremlin an empty expanse and small, unprepossessing buildings. One's gaze moves around involuntarily in search of the usual object, of what has always been there.

In earlier pictures, the hamlet of Aksakovo had an organizing center—a little white church in the middle with a square in front of it, and farther in the distance, Aksakov's house, with its buildings arranged in the form of the Greek letter π. Arranged around this, so to speak, ancient architectural complex was the rest of the hamlet. Because I didn't see the church at this time and shall never see it, and because on the square they built two stores and a dining hall, and also an oblong, light-construction House of Culture for the collective farm, the overall picture of the hamlet of Aksakovo disintegrated for me into a flat, architecturally unorganized aggregate of houses.

We arrived earlier than my traveling companions had predicted. At least three hours remained before the chairman was due to return from the session; we used this time to look over what the documents call "the memorial complex of Aksakov's estate." Of course, we started with the house, or more precisely, with the spot on which the house had stood only fifteen years before. Well, the school was just a school. The school's head of studies, Andrei Pavlovich Tovpeko, showed us around. There were tables, blackboards, hallways— everything as it should be in a new school. Could anyone object to a school, and moreover such a fine, new one? And yet, and yet, why "in place of" and not "beside"? Particularly since right here, during this excursion, Andrei Pavlovich revealed that it had been unwise to construct a school on an old foundation, that the rectangle of the old foundation restricted the school's overall dimensions, and that now its internal layout of rooms was cramped. But the school's windows open out in the same direction, and the view of the surrounding area is the same one that Seryozha Aksakov saw one hundred and seventy years ago. For this reason alone, I had to walk through the school and look out of its windows at the former park, at the river and beyond, and at bare, reddish Mount Chelyaevsk.

They installed a little square in front of the school and moreover invited a specialist from Erevan for its installation. He managed to give the square in front of the school that rather boring, formal appearance characteristic of squares in front of factory offices, bus stations, or factory dining halls. Only in place of the requisite (in such cases) Board of Honor, here in the center of the square stood three gravestones of polished granite.

As we recall, these gravestones have figured more than once in the various documents I have cited in this essay, and naturally we stopped near them. All three of them were approximately the same shape. Let me see, how can I describe them to you? . . . Well, there were three coffin-shaped objects on stone pedestals that were more horizontal and oblong than vertical. Characters had been chiseled into the front surfaces of them. Research Associate of the Provincial Museum, A. S. Popov, had been unable to read all of the inscriptions, but this time we managed to read them. Evidently they had been able to restore and clarify the letters somewhat, for they had all been worn down and were crumbling. These were the gravestones from the graves of the writer's father, Timofei Sergeevich; his mother, Maria Nikolaevna; and his brother, Arkady Timofeevich. The gravestones were arranged in a row, one beside the other, in the middle of the square in front of the school, where, according to the usual plan, one could expect a Board of Honor. Right away I asked Andrei Pavlovich Tovpeko to show me the location of the graves themselves. According to A. S. Popov, "On the spot where the church had stood (it had been built by Sergei Timofeevich's father at the end of the eighteenth and beginning of the nineteenth centuries) lay a pile of road-metal and garbage, while next to this pile three gravestones lay in disorder." It was obvious that he was talking about them, about these gravestones; it was obvious that the graves had been located next to the church, which fact Andrei Pavlovich Tovpeko confirmed for us: "A small chapel had stood near the church, and not far from it was a crypt. Sergei Timofeevich Aksakov's parents had been buried in it. Let's go to the square and I'll show you the spot."

We came to a level, asphalt-paved square that was built up on four sides with the low, silicate-brick buildings of two stores, a dining hall, and the collective farm's House of Culture. There was no longer any road-metal or garbage here. Just as there was no trace of the Znamenskaya church, which at one time had stood on this square. Only at the entrance to the House of Culture, instead of a threshold lay a large, semicircular, flat stone which didn't at all match the silicate brick and slate, and which evidently had been part of an old church building. It is possible that it had been located in front of the entrance to the altar. After stepping on it, we proceeded into the House of Culture and found ourselves in some small, bleached, light-blue, low rooms—cubicles heated to the point of stupefying stuffiness. One cubicle contained the collective farm's scanty library. We asked the young librarian which of Aksakov's books the library had. Embarrassed, she answered that they didn't have a single book by Aksakov.

"What do you mean, you don't have a single one? You really don't have a single one? Not even an inexpensive edition?"

"Not a single one."

Behind the wall one could hear some sort of loud conversation that sounded more like a radio program. It turned out that a movie theater constituted the main and greater part of the House of Culture, and that a matinee was playing there at the moment. We peeked in for five minutes. A foreign spy was running away from our intelligence agents, now jumping out of an electric train in motion, now jumping back into the electric train. Automobiles rushed by, railroad-crossing bars were lowered, and police officers talked over portable radio transmitters. In a word, it was clear that the spy wasn't going to get away.

But I still wanted to determine the location of the crypt more precisely, so Andrei Pavlovich led me into a smooth, asphalt-paved area between the House of Culture, the two stores, and the dining hall, to a small, rectangular hatchway.

"The crypt was right here."

I looked through the opening and saw that the upper portion of it had been cemented off recently. Farther into its depths nothing could be seen.

"Yes, that's right," Tovpeko repeated, looking around, "the church stood here, the parvis was here, the chapel was here, and this is the crypt."

"But if the church and chapel were torn down, why did they leave this hole in the middle of the square? For what purpose?"

"They adapted it. According to the idea, they were planning to store water here. As a fire-prevention measure. A reservoir. The chairman will even tell you that they specially dug up and equipped this reservoir. But where have you seen such reservoirs, even if only in one village or city? They adapted a crypt. And since there's never any water in it, and so far, thank God, there haven't been any fires in Aksakovo since its very founding, the stores in turn adapted this hatchway for use as a garbage bin."

"It can't be! I can't believe it. Let's ask someone right now."

A woman was walking by, a collective farm worker about fifty years old. I addressed her and began to ask where the church had been, where the chapel had been, and where the parvis had been. The woman answered my questions and pointed out the locations to within a meter's accuracy.

"And what was this?" I indicated the hole.

"They were buried here. His mother and father. Now near the school. . . . Maybe you saw . . . the stones. . . . "

"Then what is this hole used for?"

"They dump garbage from the stores into it."

My conception of the park as a huge, tangled wisp of bast was amazingly accurate. Only a few ancient, stunted linden trees formed something like a lane in one place. The entire remaining expanse was choked with thick bushes, and also enhanced by tall, grassy plants that are now dried out and prickly.

Tovpeko attempted to explain to me where the fish-ponds, the arbor, and the park's little pond had been, in which (it seemed!) there had been swans— but it was impossible to visualize any of these things now. From the park, making our way through the bushes and thorns, we walked up to the millpond that was already covered with ice. Myriad sticks and stones had been thrown onto the ice. When we were young boys, my friends, and I too, would sometimes throw things casually to see whose stick or stone, say, would slide or roll the farthest. I was also shown the place where the Aksakovs' mill, which had burned down, had stood nine years earlier.

Now what remained was for us to take a look at what had nevertheless been done to perpetuate the memory of the writer. Well, I have already mentioned the square and the three gravestones that had been placed there in a little row. At the very front of the square in 1971 (the 180th anniversary of his birth) a monument to Sergei Timofeevich had been installed. It was a large, heavy bust that rested on an even heavier base, or to put it more precisely, on a crude, rectangular, concrete ingot. Whereas the square had been entrusted to a specialist from Erevan, for some reason the monument was ordered from Georgia and was erected (there exists a detailed account of this event by Tamara Alexandrovna Lazareva) in a hurry, at night, during a cold rain, in overly soft earth and during a piercing wind. But be that as it may, the monument stands in the square.

To one side of the square, in an auxiliary building that had remained intact and had been repaired and furnished with a slate roof, a school dormitory was located. A room about fifteen square meters in area was appropriated from this dormitory and transformed into a museum devoted to Sergei Timofeevich Aksakov. A sweet girl named Galya, a Bashkir by nationality, was the sole official employee of this museum. On all the walls of the room she had diligently hung indistinct, grainy photographs (copies of copies) that had been sent from the museum at Abramtsevo near Moscow. The writer's parents. A view of the house. A view of the mill. A view of the hamlet. Rephotographed title pages of several books by Sergei Timofeevich. It goes without saying that there weren't any objects. I was especially moved by one of Galya's ideas. She had folded pieces of white paper so they would resemble the binding of books, and had written on these "book covers": Turgenev, Gogol, Tolstoi. . . . In other words, she had made imitations of books by writers who had been close friends of Aksakov throughout his life. She had arranged these "book covers" as if they were on a bookshelf.

If I understood correctly, a battle was being waged (between which parties?) to appropriate from the school dormitory, if not the entire auxiliary building, then at least one more room for the museum. Then Galya would be granted the opportunity to hang up another two dozen photographs.

. . . Meanwhile the chairman of the collective farm Rodina, Ivan Alexandrovich Markov, was due to arrive at any minute from the session of the Regional Executive Committee. I confess that I awaited this meeting with great interest. I wanted to take a look at the man who had personally torn down Aksakov's house. In the region he had been described in the most flattering terms. An excellent landlord. Fulfills all the plans. Turns in his produce on time. Builds new houses for the collective farm workers. Gave the hospital a new building that had been constructed as the office of the collective farm. Has been decorated twice: the Order of Lenin and the Order of the October Revolution. Holds the Challenge Red Banner. Has received a multitude of certificates and awards.

All of these things somehow didn't coincide with each other: a wonderful person—and suddenly he razed Aksakov's house. And what about the crypt, adapted for use as a reservoir? And the mill that had burned down, and the pond that had been neglected? And the overgrown park, and the collective farm's library that doesn't contain a single book by Aksakov?

As a starting point in evaluating this event (the liquidation of Aksakov's house), I made a speculative assumption. Only a person who had never read Aksakov could have raised a hand against his house. It couldn't be that a person who had read *A Family Chronicle* and *The Childhood Years of Bagrov the Grandson*, who had involuntarily become accustomed to that era, had become closely acquainted with the heroes of these books—i.e., with the inhabitants of the Aksakovs' house—had experienced along with Seryozha all the joys of his childhood, had viewed the surrounding environment and nature through his eyes—in short, it couldn't be that a person who had read and, it follows, loved Aksakov would raise a hand to tear down the original (original!) house of the writer.

So near and yet so far! Only fifteen years before the original house was intact and everything could still have been remedied. And now they have to appeal to Abramtsevo: could they send at least a photograph of the house or recollections of it and verbal descriptions? And everything had depended on the will of a single person, and that person had manifested ill will toward the house, and tractors had torn it apart log by log. Does this mean that this person had not read Aksakov and had acted blindly, not knowing what he was doing? Such was my speculative premise.

Just imagine my astonishment when during our conversation Ivan Alexandrovich began to drop quotations from *A Family Chronicle*, *Notes on Fishing*, and *Notes of a Rifle Hunter*! But first, of course, we greeted each other and became acquainted after the chairman got out of his car and, smiling, came over to us as we stood and waited for him in the square next to a store. It was already four o'clock in the afternoon and we hadn't eaten since that morning, so the chairman, like a genuinely attentive host, immediately turned to the subject of dinner. It turned out that dinner was already waiting for us at the house of the secretary of the Party organization. Moreover, the dinner was hot (rich, spicy cabbage soup with pork) and accompanied by "little fire"—customary appetizers created in that region. In equal quantities they pass horseradish, garlic, and ripe tomatoes through a meat-chopper. What emerges is a spicy, watery dish nicknamed "little fire." At the table it is served in a bowl and eaten with a spoon. Over the cabbage soup and this "little fire," the conversation flowed like a river. And right here Ivan Alexandrovich Markov's erudition revealed itself. Incidentally, he cleverly evaded my direct questions by avoiding direct answers.

"Yes, they allocated the resources, but at that time they didn't feel it was possible. . . . "

"Yes, roofing iron was lying there, but at that time they didn't feel it was possible. . . . "

"The house was ripe for an accident. Its attic and upper story were packed with snow, and later the snow began to melt. . . . You yourselves understand. . . . Little kids crawl in there, and before long something happens. If a heavy beam had torn loose. . . . "

"Was it really impossible to install glass in the windows so the story would not fill up with snow?"

"At that time they didn't feel it was possible? . . . Why do you keep harping on the house? You should take a look at the school we built on that spot!"

The chairman was a man of about fifty, with rust-colored hair and a reddish, seemingly freckled face that appeared well fed and even slightly self-satisfied. Things were going well, the authorities were full of praise, gave him decorations and certificates. . . . Only why have they all become such a nuisance about this Aksakov? So these landowners, nobility, lived there; should we idolize them now? And these tourists, too . . . arrive in big groups in summer, they must have nothing else to do. . . . They should all be sent to a collective farm to dig potatoes. . . .

I attributed this rather crude thinking to the chairman during the first half-hour of our acquaintance, attempting to understand his psychology and the motives for his behavior. But of course, when he himself began to fire away entire sections out of *A Family Chronicle* from memory, I had to change my opinion. The mystery became all the greater for me, to put it more mildly, of the indifference of this caretaker of the hamlet toward the memorial places linked with Aksakov, toward this entire, to use the language of the documents, memorial complex. Now the cabbage soup had already been eaten, along with the "little fire," and I nevertheless hadn't understood anything about this man's motives and actions.

I concluded that there wasn't any mystery here and that the chairman of the collective farm was by no means a malefactor, but rather a genuinely good caretaker and probably not a bad person. My contention is not categorical only because our acquaintance was too brief and I was unable to get to know this man more broadly, in a deeper way, and more fundamentally, to make a more categorical assertion concerning his human and spiritual qualities. Let's allow that he's even a very good person.

But he's the chairman of a collective farm with all the ensuing consequences, and not at all a student-enthusiast of local lore, or a guardian of antiquities, or the chairman of the local Society for the Preservation of Architectural Monuments, or a museum employee. A chairman of a collective farm is not required to possess broad and enlightened views on our national culture, and on literature in particular, especially if the matter concerns our culture and literature of the past. A potato-digger[35] is not obligated to plant flowers simultaneously. That's not her function. In a constructive sense, her work is not adapted for this task. And if she had adapted herself to it, then in all likelihood she would have done her primary work badly.

To reiterate, I don't want to insult the vast army of chairmen of collective farms who are conscientious and diligent toilers, and who, incidentally, are becoming more and more cultured and educated. It's simply a question of other functions. A collective farm receives telephone calls and papers demanding indices and figures (and this means agricultural products); in answer to these demands the chairman provides the indices and figures. Such a concept as "memorial complex" does not fit into these two opposing currents. There's no room for it to fit in. Moreover, since the fulfillment of indices and figures requires the daily efforts of both the rank-and-file collective farm workers and the chairman himself; and since these efforts do not allow any "clearance" for the pursuit of side issues like setting a park, pond, and mill in order (which could serve only a decorative function today), naturally the chairman perceives these side issues only as annoying obstacles and distractions from the basic, pressing daily concerns of the collective farm.

In order to confirm the validity of this conclusion, let's take the thought to its extreme and utilize the mathematical method of the rule of contraries. Such a method exists in mathematics for proving theorems. For example, when we want to prove the equality of two angles we say: "Let's assume that the angles are not equivalent, then. . . . " Then it becomes an absurdity and it immediately becomes obvious that these angles are equivalent. I'm simplifying it, but in principle it's correct. Thus, the rule of contraries. The question arises: Should they transfer Yasnaya Polyana[36] to the nearest collective farm for its upkeep? And what about Mikhailovskoe? Tarkhany? Muranovo? Spasskoe-Lutovinovo?[37] What would have happened if the entire memorial complex of Tolstoi's Yasnaya Polyana had been transferred to the management and, so to speak, balance sheets of the local collective farm? Since in that place, besides a park, stands the authentic house of Tolstoi. A library, antique furniture, mirrors, parquet floors, a piano, paintings, living flowers in the house—the authentic belongings of Tolstoi. After all, all of these things must be maintained in a state of complete preservation. This requires an entire staff of employees, guards, stokers, floor polishers, experts on Tolstoi, tour guides, and gardeners.

Let's allow in addition that the collective farm had made an effort and had completed all the work that needed to be done at Aksakovo; that they had found the one million rubles mentioned in the project estimate (or let's say that the province had given them this money), had rebuilt the house, set the park and pond in order, and restored the mill. And afterward? Without an entire staff of employees and specialists in museum affairs everything would very quickly have become overgrown and dilapidated once again, would soon have lost its decent appearance, and would have fallen into a state of disrepair. Without the daily and attentive upkeep of the memorial complex, demanding in its turn daily material expenditures, the matter could not have come off.

Let's agree that it's not at all the affair of a collective farm to maintain a large and bothersome memorial-literary complex every day. Then we'll be able to understand the almost instinctive desire of the chairman of the collective farm to rid himself of the Aksakov matters that had become a burden to him, to free himself of them as radically and firmly as possible. For this we can condemn Ivan Alexandrovich Markov as a person who had read Aksakov, but as the chairman of a collective farm—hardly.

In this manner, if we want to preserve, and now actually restore the Aksakov complex, we must raise the matter to the level of state and all-union importance. We must organize this memorial complex on the same level as the aforementioned: Yasnaya Polyana, Tarkhany, Spasskoe-Lutovinovo, Muranovo, and Mikhailovskoe. To this list we may add Karabikha and Polenovo, or at least the very same Abramtsevo near Moscow.

Right at this point people might say: "There's already one Aksakov complex—Abramtsevo. Isn't that enough?"

First, however, no one to date has suffered from the fact that we have three Chekhov memorial complexes. There's the House-Museum in Moscow, the House-Museum in Yalta, and the House-Museum in Melikhovo.

Second, Abramtsevo has become more of a Mamontov (Vasnetsov, Vrublyov, Serov, Polenov, Korovin) complex than one devoted exclusively to Aksakov.

Third, the main reason: Abramtsevo is located near Moscow, where there are many other museum, tourist, and excursion destinations in the vicinity. But, you know, in Buguruslan in the steppes of Orenburg the Aksakov complex would be the only one for five hundred kilometers around, the sole and much-needed hearth in support of our culture, which would attract school excursions as well as unrestricted tourist groups, and which would combine in itself elements of enlightenment, of the cultivation of love for nature in our homeland (the cultivation of patriotism), and even of relaxation. I am opposed to the construction of tourist centers near literary memorial places, but there, in the remoteness of Orenburg and, so to speak, in its "museumlessness" ["bez-muzeinost"], I could even go so far as to support the organization of a tourist center. Moreover, the beautiful pond, when cleaned out, and the little Bugu-ruslan River itself, and the park, when set in order, and also the neighboring copses would predispose visitors to a healthy and at the same time, cultural rest.

But if we feel that Aksakov as a writer and a literary-historical phenomenon is unworthy of having his memorial place raised to the same level as those of Turgenev and Tyutchev, Tolstoi and Nekrasov, Lermontov and Pushkin, Pole-nov and Chekhov, and that the hamlet of Aksakovo can be a literary monu-ment only of local significance on the balance sheets of a collective farm or region (or even of a province!), then regarding this matter we had better put an end to all the discussions, correspondence, resolutions, decrees, acts of inves-tigation, projects, and estimates. The many years' and fruitless history of dis-cussions, projects, resolutions, acts, and estimates confirms the validity of this sad conclusion.

My trip to Aksakovo clearly could not come to an end without one penetrat-ing motif linked with nature. This occurred after the train had already started to move. I was standing by the window in the passageway of the car and watching as the hills and dales raced by. Incidentally, it was still autumn, and the direct, blunt breath of winter couldn't be felt yet, but the train (the long-distance Karagandinsk) approached the Buguruslan Station with snow-cov-ered footboards, and this snow was no longer melting. Across the golden autumn lands of the Orenburg region we were carrying to Moscow on the train's footboards the fine, biting snow of the Karaganda steppes.

At this time a fellow passenger stopped next to me at another window. We were standing at two different windows but looking in the same direction.

"Aksakov's territory!" the fellow traveler informed me. "All of his hunting and fishing trips took place here."

"There were many gamebirds here, and various animals, but now their num-bers have decreased."

"The number of gamebirds and animals has decreased everywhere. It's the twentieth century. But do you know what kind of miracle occurred in Aksako-vo last year?"

"Tell me."

"A pair of swans almost established themselves on the pond in Aksakovo. They arrived in spring and remained here to raise their nestlings. What on earth could have brought them here? Perhaps some sort of distant memory. Something may have been transmitted through their . . . genes. Perhaps at one time their ancestors had been established here, and in the descendants' blood the memory of this place was awakened. But, you know, if they had raised their nestlings, those nestlings would have flown back here the following year, as to their homeland. They would have returned without fail. So you look around, and swans would have put down roots here. They would have beautified the pond and, in general, so to speak, the landscape. It's beauty, after all, when wild swans swim across a pond! And it also would have been something like a reminder of Aksakov, as one who valued and glorified nature."

My interlocutor fell silent, and after a minute or two I dared to ask him: "What about the swans? Why didn't they put down roots?"

"Why, why. . . . They laid two eggs and began to sit on them to keep them warm. But someone took these eggs away from them. Maybe it was little boys. Because adults are no fools. They were enormous eggs, bigger than those of geese. Well, they immediately flew away from the pond and no longer have been seen there. But it's a pity. . . . "

"Of course, it's a pity," I confirmed. "Is it such a bad thing to have swans on a pond? You know, I heard that in other countries—Poland, Czechoslovakia, and Germany—it seems that swans simply swim on the lakes. There are people all around, the inhabitants, and they swim with their cygnets."

"I'll say this: we ourselves are at fault. Because of our behavior we are clearly unworthy of having swans swim in our country. We haven't deserved them. Brother, swans are something you have to deserve. . . . "

The train kept going, and it quickly grew dark. It was time to go to the dining car to have supper. My trip to Aksakov's territory was drawing to an end.

GREATER SHAKHMATOVO

I

HOW CAREFREE IT made me feel to write the previous essays about Derzhavin's Zvanka on the bank of the Volkhov River and the hamlet of Aksakovo on the bank of the Great Buguruslan River! In the words of the ancients, this really was a *tabula rasa*—a clean board. After all, they wrote with pointed sticks on wooden boards covered with wax; and so, one could easily erase or smooth over what had been written previously. What emerged was a clean board, a *tabula rasa*, a clean page, as one might still describe it in more modern terms; no matter what one scribbled, it was entirely one's own, everything appeared as if written for the first time.

Where is this Volkhov? Where is the Buguruslan? Can many writers, and moreover those from Moscow, say that they have set foot in and trodden through the ruins of Derzhavin's house, and likewise through the place where as recently as 1960 (!) had stood the solid house of Aksakov, which had been built in the eighteenth century?

Aside from all this, what was the point to the preceding essays? To attract people's attention, to awaken their interest, and to demonstrate the necessity of restoring these memorable literary places that are dear (as one would like to believe) to the heart of every compatriot.

Even if a snowball is as big as a hill, it starts out about the size of an apple. As children, we would sometimes form such a snowball and let it roll down a steep hill under favorable conditions, which consisted of a thaw and damp, sticky, freshly fallen snow. And soon, in place of an apple, there would be a cap, and then a sphere the size of a wagon wheel, and then a mass taller than we ourselves.

No, my preceding essays about Derzhavin and Aksakov didn't become the kind of snowballs that turn into boulders of snow. Perhaps they didn't turn out the way I had intended, or perhaps the external circumstances were unsuitable, but at any rate, nothing stuck to my snowballs. No public interest was aroused, no continuation of the theme appeared in the press, nothing was heard about them even in conversations or judgments, and no one is planning to restore Derzhavin's Zvanka and Aksakov's house in the Buguruslan District of Orenburg Province.

To be sure, there were readers' responses, and many of them, but you have to agree that, after all, you can't build or revive a house, much less an eighteenth-century country estate, out of readers' responses.

Let me reiterate (this is also important for the present essay) that examples of complete restoration and revival of ruins exist, and they are striking ones, so the idea of rebuilding and restoring something on a bare piece of land is not a fantasy. Moreover (and we must especially understand this), restoration on a bare piece of land doesn't signify at all the restoration of something out of a bare piece of land, out of nothing. The fact remains that magnificent ruins, as it were, continue to exist: they seem alive and eternal, and we need only to materialize them, to give them once again a material (boards, nails, plaster, roofing iron, glazed windows) aspect.

And there have been such instances. There are dozens of examples of newly restored memorial objects that had virtually been worn down by time or destroyed in other ways.

Pushkin's house in Mikhailovskoe burned down in 1919 and was rebuilt again in 1937 for the centennial of the great poet's death. The house burned down a second time during the war and was restored in 1949.

Lermontov's house in Tarkhany was also destroyed by fire and rebuilt.

Similarly lost and rebuilt are the respective houses of Repin in Penaty, of Chekhov in Melikhov, of Turgenev in Spasskoe-Lutovinovo. . . . One can find many examples if one looks for them in our immense country.

The Triumphal Arch was completely restored, albeit in a new location: it is no longer near Belorussky Station, but at the end of Kutuzov Avenue.

My purpose in writing these essays, I repeat, was to awaken interest and attract attention. And if suddenly there were to appear in a central newspaper, and not just in any newspaper but in *Pravda*, a short notice like the one that follows, this would justify a writer's efforts, this would be his reward, this would constitute the desired result. A little notice in *Pravda*, March 7, 1977, on the last page under the column "Projects." The notice is entitled "The House with a Mezzanine Will Rise."[1] The text, signed by N. Dorofeev, reads as follows:

> The All-Union Scientific-Restoration Enterprises, under the Ministry of Culture of the USSR, have developed a project to restore the country estate Shakhmatovo, where the poet Alexander Blok spent his childhood and youth.
>
> In one of the scenic places near Moscow not far from Solnechnogorsk, the estate of Shakhmatovo was located at one time. A copse of trees with a girth as large as one's arms can encircle now grows on the spot where, sixty years ago, a house with a mezzanine had stood.
>
> The architects V. Yakubenya and A. Chekhovskoi—the authors of the reconstruction project—studied the poet's early poems, trying to find some descriptions of the estate. At first the work did not progress well. There weren't enough photographs. It's true that there were drawings in Blok's hand as a child, but as a rule they depicted only a part of this or that structure: of a house, outbuilding, or bathhouse. But they found something that helped them. . . .
>
> In the archives of Pushkin House in Leningrad, a family chronicle was found that had been compiled by the sister of Blok's mother, Maria Beketova. Among reminiscences about the Blok and Beketov families a detailed description of Shakhmatovo was discovered, and it described not only the main house, but the various outbuildings as well, tracing the history of their construction. Even the plan for the internal arrangement of rooms and furniture had been preserved. Everything was mentioned, down to the pattern and color of the wallpaper. . . .
>
> In summer of last year a group of workers conducted excavations at Shakhmatovo on the spot where the main house had stood. They unearthed some remains of tiles, half-rotted pieces of wood that at first glance didn't appear to be anything of interest. But the restorers had become imbued with the spirit of Shakhmatovo, "the house with a mezzanine," to such an extent that in these wooden objects they recognized parts of the original casings and frames.
>
> Work on the reconstruction is scheduled to begin in summer.

A letter written to me on the same subject nearly coincided in time with the above notice:

> Dear Vladimir Alexeevich:
>
> I am sending you a recently discovered photograph (from the 1920s) of the Church of Archangel Michael (of the eighteenth century) in the village of Tarakanovo, where Alexander Blok and Lyubov Dmitrievna Mendeleeva were married, and where a burial service was read over A. N. Beketov. At present, architects of the All-Union Scientific-Restoration Enterprises, under the Ministry of Culture of the USSR, are developing projects to restore the church in Tarakanovo and the house at Shakhmatovo. However, this doesn't yet predetermine

the realization of the projects, especially as the centennial of Alexander Blok will take place in 1980. It is desirable to try to include places associated with Blok that are auspiciously located on the road between the two capital cities[2] among the ranks of so-called objects of Olympic display. At the very least, a concert hall could be housed in the church, while a museum could be installed in the house. In any case, feast your eyes, see how beautiful the church at Tarakanovo was, the church of a country estate, built with care. I wish you all the best.

With sincere respect,
S. Lesnevsky

Although a skeptical tone occasionally appears in the letter ("this doesn't yet predetermine the realization of the projects"), and although it's not clear why the house of the great Russian poet Blok should be included, in order to restore it, among the ranks of the so-called objects of Olympic display, it still represents some kind of movement and hope. If only a similar notice about Derzhavin's house or Aksakov's house were to appear in *Pravda*!

The forward motion concerning Shakhmatovo is picking up; interest in it has been awakened, and attention has been drawn to it. Since 1970 annual festivals of poetry have been held at Shakhmatovo at the beginning of August. If one believes an official write-up, among those who have participated in the festival are the following writers and poets: Pavel Antokolsky, Alexei Surkov, Konstantin Simonov, Irakly Andronikov, Lev Oshanin, Evgeny Dolmatovsky, Margarita Aligér, Sergei Vasiliev, Vladimir Soloukhin, Ludmila Tatyanicheva, Alexander Mikhailov, Viktor Bokov, Mikhail Lvov, Ekaterina Shevelyova, Rimma Kazakova, Larisa Vasilieva, Vladimir Sokolov, Evgeny Evtushenko. . . . Those who attended the festival include Sergei Konenkov, Mariètta Shaginyan, Rasul Gamzatov, Sergei Orlov, Boris Slutsky, David Samoilov. . . . The festival was discussed by the local population, the general public, and the press, including the following newspapers: *Pravda*, *Izvestiya*, *Sovetskaya kultura*, *Komsomolskaya pravda*, *Literaturnaya gazeta*, *Literaturnaya Rossiya*, and *Leninskoe znamya*.

And so, is it worth our time to make a new snowball, when it would be easier to attach ourselves in the form of an additional handful of snow to what is already rolling downhill like an avalanche?

But even though the snowball is rolling, things are not getting done. In March they published that "reconstruction work is expected to proceed in summer"; however, I'm writing these lines in late fall, while in Shakhmatovo not a single bitter burdock or stinging nettle has been moved. So at the moment there is no end in sight to this matter, and hence let's begin now, if I may once again refer to the laconic, expressive piece of wisdom of the ancients—*ab ovo*—from the egg.

The mind, albeit an analytical one, cannot always grasp reasons and consequences. One can emphasize, but only with a small grain of paradoxicalness, that if the great Russian scientist Mendeleev had not purchased an estate in Boblovo in 1865, we would not have had the poet Blok, at least not the way we know him. And it's possible that the poet in him might not have existed at all.

The chain of natural laws in this matter is as follows. Dmitry Ivanovich Mendeleev advised his friend—the prominent botanist, professor, and (later) Rector of St. Petersburg University, Andrei Nikolaevich Beketov—to acquire the small estate of Shakhmatovo, which was located near Boblovo. And Beketov acquired it in 1874.[3] During the summer months the large Beketov family would gather at the small estate. In 1880 the professor's daughter, Alexandra Andreevna, gave birth to a son, Sasha, who at the age of six months was brought to Shakhmatovo from St. Petersburg in the summer of 1881, during the time of luxuriant blooming of grasses and flower gardens.

In my mind it was like this: the warmth of summer with its blue sky and white clouds, its pink clover and bright-green fields of rye (the little boy was six months old in the second half of May), its clumps of hundred-year-old lilacs and parterres of dog-rose bushes, the glow of the evening sky and fragrant silence, the buzzing of bees and fluttering of butterflies—Blok was plunged into all of these things in the heart of Russia as into a font; this was in truth like his second baptism: a baptism by Russia, the Russian outdoors, the Russian village, Rus. The baptism was purely symbolic, since a six-month-old child couldn't see, feel, or understand very much; but, after all, later on he was brought to this place again and again every year, to this paradisiacal little corner (". . . a corner of paradise not very far from Moscow," in the later description of Blok himself just before his death), and already he was one and a half years old for his second visit, two and a half years old, three and a half years old. . . .

> I sank into a sea of clover,
> Surrounded by the stories of bees.
> But my childish heart was discovered
> By the north wind that beckoned to me.[4]

These lines are from the year 1903. Blok was already twenty-three years old. He was already a marvelous poet. But this is really firsthand evidence of the fact that if there comes a moment when in someone's heart and soul the spark of God, as they say, is initially ignited, when a person is struck by a heavenly arrow, a lightning flash of talent and purpose, then for Blok this lightning struck, and this spark flared up, precisely at Shakhmatovo, when "[He] sank into a sea of clover, / Surrounded by the stories of bees." Without any exaggeration we can now state that if Blok the man was born in St. Petersburg in a rector's house, Blok the poet was born at Shakhmatovo.

To be sure, also significant are genes and their particular combination in order for Pushkin and Lermontov, Tolstoi and Esenin, Nekrasov and Blok to develop in the soil of their native culture; there must be readily available combustible material that will flare up from the spark dropped into the soul; but the spark is also a vital element. It could come from a specific condition in nature, some kind of unusual illumination, a song that was heard, a gentle spring rain, the morning dew, the evening bells. . . .

Here is a prosaic but well-documented and well-known example. When one of the greatest contemporary scientists, an academician, was a boy, in his presence a pig was butchered unsuccessfully. Evidently the butchering went on for a long time, with heart-rending squeals and the gushing of blood. Maybe the pig even broke loose and dashed about with the butcher's knife plunged into it. The sight affected the boy to such a degree that it consequently determined his later entire scholarly activity: for the rest of his life he was preoccupied with the invention of artificial albuminous food that could replace in the human diet the meat of living creatures, including even fish.

But if the confluence of negative conditions, the clot of negative emotions, if (let's say) a dark flash of lightning can strike the soul and rearrange many things in it, then why can't the same effect be produced by a bright, life-giving flash of lightning that issues from the confluence of positive conditions? And if combustible material were already in reserve, then the flame of poetry would immediately be ignited:

> I sank into a sea of clover,
> Surrounded by the stories of bees.
> But my childish heart was discovered
> By the north wind that beckoned to me.

Let's take a look at this place, Shakhmatovo, in more detail, bearing in mind that in order to visualize the Shakhmatovo of those times, we shall have to see it through the eyes of its contemporaries, its eyewitnesses. But the eyewitnesses were observant and talented in their own way, and in this sense we are fortunate. Several eyewitnesses had looked at Shakhmatovo even before Blok.

But first—some objective information. In Pushkin House in Leningrad, in the so-called A. Blok Collection, a map of Shakhmatovo is housed, written in the calligraphic handwriting that at one time took the place of our typescript.

Map of the estate Shakhmatovo, of Moscow Province, Klinsk District, Vertlinsk Volost,[5] property of the heiresses of the Privy Councillor A. N. Beketov: Sofya Andreevna Kublitskaya-Piottukh, wife of the active councillor of State; Alexandra Andreevna Kublitskaya-Piottukh, wife of the colonel; and the noblewoman Maria Andreevna Beketova. Photograph taken in 1908. Scale: 100 sazhens[6] = 1 inch.

On the map the Shakhmatovo lands are highlighted with various forms of shading and embellished with different colors; surrounding the lands on the map is a white field containing the following places: "Land of the peasants of the village 'Osinki,'" "Land of the peasants of the village 'Fomitsyno,'" "Land of Mr. Batyushkov," "Land of the merchant Zaraisky," "The Eskinoprasolovsk Official Forest Dacha.". . .

The shaded and embellished spot of Shakhmatovo itself is subdivided as follows (my citation is in a simplified form [1 desyatina = 2.7 acres]):

1.	Planted with deciduous trees (aspen and birch)	71.00 desyatinas
2.	Planted with deciduous trees (alder and aspen)	17.00 desyatinas
3.	Cleared land	1.50 desyatinas
4.	Spruce plantation	1.80 desyatinas
5.	Pasture	0.60 desyatinas
6.	Hayfields	16.00 desyatinas
7.	Clover fields	3.00 desyatinas
8.	Ploughed fields	0.60 desyatinas
9.	The estate	2.45 desyatinas
10.	Swampland	0.90 desyatinas
11.	Land adjacent to roads and streams	2.00 desyatinas
	Total:	116.85 desyatinas of land

If we eliminate the forests that don't require seasonal agricultural work but only maintenance, as well as the streams, pasture, cleared land, swamps, roads, and also the estate itself—the garden, flower beds, pond, all the hundred-year-old lilacs, dog roses, annex, lanes, the clearing in front of the house, little paths and secluded nooks—there remain twenty-five desyatinas of agricultural land proper, the meadows and fields. Heaven knows, it's not that much, if we keep in mind that Tolstoi, for example, aside from his Yasnaya Polyana properties, purchased land, somewhere near the Volga and somewhere in Samarsk Province, in thousands and tens of thousands of desyatinas.

Strictly speaking, only the Beketovs themselves wanted, according to nobility's traditions, to call Shakhmatovo an "estate," a "country estate," a "small hamlet." It was really just a dacha, a summer cottage with a little bit of pleasant land.

Of course, they could establish some kind of household with the few cows and horses they had there; they could sow some oats, rye, and clover. It seems to me that the economic value of three desyatins of clover was insignificant in comparison to how it, the clover field, once appeared to a child enchanted by a clover field with the "stories" of golden bees.

The Beketovs and Bloks were not landowners, but working members of the intelligentsia.

Andrei Nikolaevich was a professor, Rector of St. Petersburg University, founder of courses on Bestuzhev, and the teacher of Timiryazev, the future scientist. Blok writes about the professor's wife—that is, about his own grandmother—in the following words:

My grandfather's wife, my grandmother Elizaveta Grigorievna, was the daughter of the well-known traveler and explorer of central Asia, Grigory Silych Korelin [also not a "parasite"—V.S.]. All her life she worked on the compilations and translations of scientific and literary works; the list of her works is enormous—during her last years she translated up to two hundred typed pages per year. . . . She was very well read and knew several languages. . . . She translated many works of Buckle, Brehm, Darwin, Huxley, Moore (the poem *Lalla-Rookh*), Beecher Stowe, Goldsmith, Stanley, Thackeray, Dickens, W. Scott, Bret Harte, George Sand, Balzac, V. Hugo, Flaubert, Maupassant, Rousseau, and Lesage. This list of authors is far from complete. . . . Payment for her work was always paltry. Nowadays these hundreds of thousands of volumes in inexpensive editions are out of print, and those acquainted with the prices of antiquities know these days how expensive even the so-called 144 volumes (printed by a Mr. Panteleev)[7] are, in which many translations by E. G. Beketova and her daughters are included. A characteristic page in the history of Russian enlightenment. She personally knew many of our writers, and occasionally met with Gogol, the Dostoevsky brothers, Apollon Grigoriev, Tolstoi, Polonsky, and Maikov. . . . I cherish the copy of the English novel that F. M. Dostoevsky personally gave her to translate; the translation was published in *Vremya*. . . . Their daughters—my mother and her two sisters—inherited from these grandparents a love for literature and an unsullied understanding of its lofty purpose. All three daughters produced translations from foreign languages. The eldest, Ekaterina Andreevna (by marriage Krasnova), was well known. Two of her own books of stories and poetry (the last of which received the prestigious prize of the Academy of Sciences) were published posthumously. Her original *nouvelle, It's Not Fate*, appeared in *Vestnik Evropy*. She translated from French (Montesquieu, Bernardin de Saint-Pierre) and Spanish (Esperonceda, Bécquer, Peres Galdos); and rewrote English tales for children (Stevenson, Haggard; published by Suvorin in the "inexpensive library").

My mother, Alexandra Andreevna . . . translated and still translates poetry and prose from the French (Balzac, V. Hugo, Flaubert, Zola, Musset, Baudelaire, Verlaine, and Richepin).

Maria Andreevna Beketova translated and still translates from Polish (Sienkiewicz and many others), German (Hoffmann), and French (Balzac, Musset). She produced new editions of popular works (Jules Verne, Silvio Pellico), biographies (Andersen), and popular monographs (*Holland, The History of England*, etc.). Not long ago Musset's *Carmosine* was performed in her translation in a workers' theater.

We don't even have to consider whether or not Blok himself worked: he created an entire epoch in Russian poetry and left behind a collected works in many volumes.

Nowadays, if some Moscow writer translates one or two novels or two or three *nouvelles*, she[8] quickly applies to the Writers' Union, to the translators' section, and she also applies to Litfond and they accept her, and no one would consider her (perhaps rightly so) an idler, or for that matter an exploiter, and upon becoming a member of the Writers' Union, her conscience will always be clear.

But these toilers, I would say, grand toilers in the field of Russian enlightenment, felt anguish and pangs of conscience, since right beside them peasants from nearby villages were raking hay, ploughing the earth, and carrying sheaves of grain.

Recorded in the family chronicle as an extraordinary occurrence is an episode that became a stimulus for Blok to write the essay "Neither Dreams, Nor Reality."[9] I call it a stimulus because in the essay everything is expanded into a symbol. And somehow the episode itself seems rather innocuous. . . . Nevertheless, here are the words of Maria Andreevna Beketova (*A Family Chronicle*):

> I'll never forget the time when we were eating dinner on the balcony on a fine summer day. It was haymaking time. The entire yard had been mowed, and the peasant men and women had come to clear away the hay that lay right near us. At this moment I experienced a feeling of burning shame for the quantity of delicious food before us, our servant, and our entire domestic setting. . . . These people, who worked all day on our behalf for a few kopecks. . . .

Let's clarify that the Beketovs didn't have their own peasants, their own *muzhiks* (male peasants). And, of course, no one owned any at that time. Serfdom had been abolished long before Professor Beketov had purchased Shakhmatovo with its land. One could only hire peasants to work, and the latter could either agree to do the work or not agree. The payment was mutually agreed upon. The prices for different kinds of work were most likely settled upon and fixed. They were probably not high—say, fifty kopecks per day for each person, but remember that twenty-five rubles were already enough for a cow. And here these people, these women, each of whom translated two hundred typed pages per year, felt ashamed that peasants raking up hay had come upon them during their dinner on the balcony! Their neighbor from Boblovo, Dmitry Ivanovich Mendeleev, on occasion would dine with them. Should one refer to him, too, as an idler and sponger? One also might encounter Blok's stepfather, Colonel Kublitsky-Piottukh, at the table. And although he wasn't the discoverer of the periodic table of the elements, he still was a commander of a brigade. Is this easy work?

But let's return to Shakhmatovo. In Maria Andreevna Beketova's words:

It's a small estate located in the Klinsk District of Moscow Province; father bought it in the 1870s. The countryside where it's situated is among the most picturesque places in central Russia. The so-called Alaunsk Hills are located here. The surrounding environs are hilly and woody. From some of the higher elevations one can see for miles around. Father was attracted to Shakhmatovo precisely because of its beautiful views of distant areas, the charming countryside surrounding it, and also its coziness, which is well suited to life on an estate. The old house with a mezzanine was not large, but it was sturdy; rooms that were laid out so cozily also contained antique furniture, and there were even some pots and pans in the kitchen. All of the outbuildings turned out to be in good condition; in the coach-house stood a carriage on springs and a troika of healthy, dun-colored horses; there were also cows, chickens, and ducks—all for the future owner's use.

The nearest postal station, Podsolnechnaya, which included a large trading village and the hospital of the zemstvo,[10] was eighteen versts from Shakhmatovo. To get there one had to take a country road that, near Shakhmatovo, had potholes and ruts in it and led through the magnificent state-owned forest, "Praslovo." This forest stretched for many versts, and bordered our land on one side. Our country estate . . . stood on a high hill. The drive to the house led through a spacious yard containing rounded flower beds of dog-rose bushes. A shady garden with old linden trees was located on the southeast side, on the other side of the house. Upon opening the glass door of the dining room, whose windows looked out onto the garden, and stepping out onto the terrace, everyone was struck by the spacious and varied view. . . . In front of the house was a little sandy square with flower beds, and behind the square some ancient, spreading lindens and two tall pines formed a stand. In the garden were bird-cherry trees, a multitude of lilacs, white and pink roses, and a dense, semicircular bed of white narcissus and a similar bed of purple irises. One of the side paths had a canopy of very old birches, and it led to the gate that opened onto a lane of spruce trees descending sharply to the pond. The pond was in a narrow valley with a stream running through it; the stream had a canopy of enormous spruces, birches, and young alders.

Such was this enchanting place. . . .

The quotation is taken from Maria Andreevna Beketova's book entitled *Alexander Blok: A Biographical Sketch.*[11] It was published in 1922 by the publisher Alkonost. It is a rare book and an enormously interesting one. But Maria Andreevna wrote another work, entitled *Shakhmatovo and Its Environs,* better known to Blok scholars by the title *A Family Chronicle.*[12] The manuscript of it is housed in Pushkin House in Leningrad, in the M. A. Beketova archive of the Blok Collection; it bears the note, "An accurate copy of the manuscript with corrections by an unknown person. 1930." Someone rewrote, recopied this manuscript, already using the new orthography without the letter *yat*[13] and the hard sign at the end of some words. But the paper is still old, prerevolutionary, large-format, heavy, and lined with pink and black vertical and horizontal lines. In the past, bookkeeping and office records were kept on this kind of paper. Anyone who has ever seen this paper can picture it; one cannot describe it to those who haven't.

The year is 1930. This means that at death's door the toiler Maria Andreevna was preoccupied with making the manuscript legible, and hence some Leningrad woman among her younger or dearer friends spent many weeks scratching with a quill pen in order to recopy this marvelous document accurately.

I cannot understand why this manuscript has not yet been published, why it hasn't even appeared in some professional journal for literary scholars.

The chapters of the manuscript include "Everyday Life at Shakhmatovo," "Original Appearance and Furnishings of the House," "Shakhmatovo's Front Yard and Garden," "Walks at Shakhmatovo," and "Neighboring Villages and Peasants.". . .

To be sure, all of this is still mainly a description of the Shakhmatovo estate before Blok's time, but nevertheless it's precisely the setting in which Blok found himself when he made his appearance in the world and became yet another resident at Shakhmatovo. So this description of the estate "before Blok" doesn't evoke any feelings of annoyance, except for one passage—the description of the library: some things could have changed in Blok's time, even the house was repaired and rebuilt (in 1910), and the annex was reequipped. Blok loved to cut down old trees, plant roses, and create flower beds—he enjoyed this kind of activity. But these changes are not fundamental; the library, however, may have changed radically. In Maria Andreevna's description it looks like this:

I must say that our furnishings and dress at that time were very modest. . . . But for all that, we had many books: we always subscribed to one or two Russian journals and to *La Revue des Deux Mondes*, and Father would order additional scientific journals, most of them German. Of course, he had an entire library of scientific books in four languages. Father's literary library, which was far from being complete, consisted of the Russian classics of poetry and prose. With time the collection grew. It also included Goethe's *Faust* and Shakespeare in Russian translation. In the original he had the complete Schiller, *Faust, The Book of Songs* by Heine, almost all the novels by Dumas, and the plays of Alfred de Musset. I cannot remember everything that he had. But I repeat that there were many books. . . . At Shakhmatovo there gradually evolved a small library of old journals (*Vestnik Evropy, Vestnik inostrannoi literatury, Otechestvennye zapiski, Severnyi vestnik, La Revue des Deux Mondes*, and others), classics, literary collections, and small pale-yellow tomes of Tauchnitz; Mother and my elder sisters loved to read the last of these, i.e., English novels. All of these books were kept in various rooms on bookshelves, since there were no bookcases at Shakhmatovo. In addition, music was sometimes brought to Shakhmatovo: the sonatas of Beethoven, Haydn, and Mozart; many songs by Schubert for singing; several waltzes and ballads of Chopin; *Songs Without Words* by Mendelssohn; and operas transcribed for the piano, including Mozart's *Don Giovanni* and Gounod's *Faust*. In addition, there were three more bulky, handwritten notebooks in which Mother, while still young and without the money to buy notes, copied in her unusually clear and steady handwriting Beethoven's sonatas, many excerpts from Meyerbeer's opera *Robert le Diable* and from other operas, Schumann's *Manfred* in its entirety for four hands, and myriad old romances and other pieces. In one of these books in a red binding were some printed notes, too: romances by Varlamov, Gurilyov, etc. . . .

Later on in Maria Andreevna's text one encounters the sentence, "All of this, unfortunately, perished along with everything that was in the Shakhmatovo house." But let's not get ahead of ourselves. I think that if Shakhmatovo had been intact during the years when Maria Andreevna was writing *A Family Chronicle*, she would not have described it in such detail. One remembers what has been lost with especial clarity and pain.

My sister Katya would always wipe her toilette table under the mirror with white muslin that had two ruffles along its upper and lower edges, and she would arrange in a pleasing way the various small items on it, such as bottles of perfume, powder-cases, little vases, etc.

. . . [The storeroom] didn't take up much space, approximately three square arshins. Along the walls there were shelves, on which boxes of provisions were kept: they were arranged in small groups, some containing spices . . . all of them were labeled. In spring they would be spread out on the balcony to dry in the sun. Special boxes were ordered for white flour and granulated sugar. We purchased granular wheat flour in large, five-pood[14] sacks and granulated sugar by the pood, since, in addition to desserts, it was needed for preserves. Sugar for tea and coffee was purchased by the loaf, and Mother would break it mostly herself, using a special utensil with a heavy, hinged knife that was attached to a lower drawer. Tea and coffee were always brought from St. Petersburg, while the remaining provisions came from the postal station. . . . From St. Petersburg we would also obtain the best olive oil for salads. . . . Steelyards hung suspended from nails . . . the storeroom was locked with a padlock. . . . During our meals the conversation was a general one and very animated. We talked about various things: household matters, politics, literature. . . . A day at Shakhmatovo was organized just like in the city: morning tea, breakfast at one o'clock P.M., dinner at six, and evening tea around ten; there was no supper. . . . At tea, at a table covered with a white tablecloth, one would find Mother, dressed in a roomy housecoat of light-colored calico and a comb covered with black lace on her head, pouring tea out of a large, brass samovar. . . . On the table there were homemade rolls, fresh creamery butter, and cream. . . . Father drank tea out of a special cup; his tea was very strong and sweetened with a little spoonful of homemade black-currant preserves that were served in a small painted dish brought from the Trinity–St. Sergius Lavra. . . . Great importance was attached to gravies and sauces, especially the latter. With boiled chicken and rice (cooked from the best kind until it was soft, but without fail the kind that separated rather than forming clumps) they served a white butter sauce [this wouldn't be written in any other way in the 1920s!—V.S.], and lemon, lightly fried and floured; for fried meat they often made a sauce with marinated saffron milk caps[15] [very likely, even at the beginning of the 1920s!—V.S.] . . . in great demand were such dishes as soufflé of fish or game, always accompanied by special sauces . . . fowl was sliced into long, thin pieces [was all this written in 1919?—V.S.], not hacked across its bones. . . . Meat was sliced into thin pieces, and without fail along its fibers. . . . We always hired good, discriminating cooks, but typically, despite the great humanity and even kindness of the landlord and landlady, it wouldn't have occurred to anyone that a late dinner in summer forced the cook on hot days to spend the entire day sweating in the kitchen, and on the whole to have little free time. To be sure, she always had a dishwasher at her side, so that she was spared the task of washing an entire pile of dishes—but no one thought of sparing the dishwasher, either. The servant received excellent food and was treated well, but the cook was inundated with work. Sometimes we had three dishes made out of dough in a single day, for example, fruit dumplings for breakfast, pirogies for dinner, and buns for evening tea. It was easier for the maid, especially as we hired a laundress separately. But,

91

even so, one might add that because of the good food and country air at Shakhmatovo a servant always became healthier and was usually merry. In our family the cook was considered a very important person, since we attached great importance to good food. . . .

. . . Let's turn our attention to the house. It had one story with a mezzanine, and was built in the style of the typical country estate of the 1820s or 1830s. Cozy and well-situated, it was built of splendid pine on a brick foundation, with gray wooden trim and a green iron roof. A kitchen was built near the house and was linked with it by an enclosed outer hall. . . . The house consisted of seven habitable rooms—five downstairs and two on the mezzanine. . . .

What follows is a detailed description of the rooms, their dimensions, the staircases, landings, stoves, passageways, cloakrooms, windows, and also the views from them, the color of the wallpaper in different rooms, the furniture and its arrangement, right down to the chairs, the purpose of each room, and their names. The Corner Room, Light-Blue Room, Red Room, White Hall, the Room Above the Door. . . .

In the dining room Mother put in the east corner a large, antique icon of the Mother of God that was set in a gilded framework; in the other rooms hung little icons or crosses. For herself Mother chose the shadiest room, which was shaded by two large, silvery poplars that stood by the fence and now stand behind the gate leading from the yard into the garden. Everyone who came from that direction passed by the window of Mother's room. . . . Mother had a simple washing-table of village craftsmanship that was covered with oilcloth; opposite the bed a mirror framed in a reddish wood stood on a pier-glass table. . . . A beautiful walnut table of polished wood with a drawer and figured footrest served as a writing table and stood with its side to the window, at which hung an old cotton curtain with bouquets of white flowers scattered on a light-gray background. In the corner on a set square of reddish wood stood an icon of the Kaluga Mother of God, in front of which a green icon-lamp burned all night. . . .

Of course, all of these ash closets, sofas upholstered in chintz, buffets, pianos, beds, curtains, lamps, chairs, love seats, toilette tables, ottomans, washing-pitchers, chests of drawers, mirrors, small round tables, stained glass, wallpaper—all of these details of everyday life at Shakhmatovo wouldn't be needed and would even be superfluous if Maria Andreevna were simply writing a biography of her nephew (they are omitted in her biography of Blok), but Maria Andreevna evidently understood that Shakhmatovo remained and existed only in her memories, and nowhere else. This realization sharpened her memories to the point of pain. Moreover, it is possible that she also understood that she was not writing an entertaining work of belles-lettres, but a document. And how good it is that now this document exists! It's one thing that we gain a complete picture of the Shakhmatovo house from it, but another thing altogether that it will be priceless if things reach the point of restoring Shakhmatovo.

And it's not just the house! The entire estate, fences, gates, outbuildings, village fences, flower beds, garden plants, farmyard, ice-house, coach-house. . . . And how all of this was arranged and situated, and how it appeared—Maria Andreevna described absolutely everything in her chronicle. For example:

> I begin my description of the outbuildings with the barn. It was built in a correct symmetrical form with a steep, red plank roof and a semicircular arc above the entrance door. Along both sides of the barn were completely identical small, low sheds with sloping roofs that merged with the barn's roof: various tools and boards were stored in one of them, while wood was stacked, and a watchdog lived, in the other. Sturdy oak granaries were located inside the barn. . . .

When one reads the chronicle of Shakhmatovo by M. A. Beketova that is kept in the museum foundation in manuscript form, and realizes that few people thus far have had access to it, one is tempted to copy as much as possible out of it. But one possesses a feeling of restraint that has been formed over a period of decades, and it dictates its own laws. Let's limit ourselves to several more lines that refer, no longer to wallpaper and dinner food, or to closets and barns, but to the green attire of the estate, its earthly beauty.

> The entire expanse of the yard not occupied by buildings and flower beds was covered with grass . . . there were two young, silvery poplars with two long benches standing beneath them; we would sit on these benches when we were expecting guests, since from this spot we could see Podsolnechnaya Road, and even farther in the distance we could hear the bells of approaching troikas. . . . In general, Shakhmatovo was noted for its gay and cozy character, which can be explained by the fact that it was located on a hill, while the garden faced the southeast. . . . Along both sides of the balcony under the windows there grew two huge jasmine bushes: their dark verdure stood out beautifully against the gray color of the house, and when they were in bloom they shone in their whiteness and smelled fragrant in the midst of the humming of fluffy bees. . . . An entire thicket of pink dog-rose bushes approached the left edge of the square . . . a wall of acacias rose up. . . . Father planted wonderful irises, white narcissus, and parterres of Provence roses in the garden. . . . Scattered here and there on the lawn were berry bushes and apple and cherry trees that made our garden indescribably

beautiful when it was in bloom. . . . Three kinds of lilacs constituted one of the main ornaments of the garden. . . . The best weeping birch of the entire garden stood on the lawn. . . . A mountain ash grew off by itself, and hence spread its branches especially wide: they started growing so low on the tree that it was comfortable to sit on them. . . . We liked our garden very much, and found in it a thousand kinds of happiness. It was pleasant just to stroll in the garden, and great fun to pick flowers in order to create countless bouquets out of the garden and field flowers. We passionately hunted for white mushrooms, and there were many of them, especially under the fir trees. . . . A multitude of songbirds could be found in the garden. Nightingales sang energetically right near the house in the dog-rose and lilac bushes, while entire choruses of them rang out from beyond the pond. Orioles liked to visit our linden trees in summer on sunny days. They filled the garden with their clear whistling, and their bright yellow color flashed as they flew from one tree to another. All kinds of thrushes could be found in great numbers. . . . There were squirrels right in the yard, and they would come from the neighboring forests to visit us, lured by pine cones and nuts. . . . Owls would appear at twilight and during the night . . . one could only see them in flight or sitting motionless on the roof of some building. . . . And what views one could see from the windows or the various corners of the garden! . . . Not for nothing did Blok call our estate "fragrant seclusion."

We lived a very secluded life. Even the nearest village was farther away than usual, more than a verst in distance, while the forests that approached us from various sides further intensified the impression of the seclusion and isolation of our summer retreat.

II

This was the kind of paradise in which Blok found himself so soon after entering the world. Starting from the age of six months, annually during the course of thirty-five years, excepting only the last five years of his life (1916–21), Blok traveled to Shakhmatovo for the summer months.

Many people consider Blok purely, or first and foremost, a "Petersburg" poet. And indeed, in reading this poet's work, city motifs emerge before anything else. Starting with the famous (and I would even say, "notorious") "Night, street, streetlamp, drugstore," with "Woman Stranger," "The only hope left for me / is to look in the well in the yard," "I'm nailed to the bar in the tavern." "Eternity cast a pewter sunset / into the city," "In taverns, on side-streets, in twists and turns, / in electric sleep when I'm awake," "Columns once again covered with snow, / Elagin Bridge, and two lights," "I sent you a black rose in a goblet / of wine as golden as the sky,"[16]—starting with all of these Petersburg motifs (and one doesn't even have to search for them in Blok's books, one only has to open them), and ending with the most "Petersburgian" poem (no matter which others one compares to it), *The Twelve*[17]—the city and its image are everywhere, in some aspects wonderful, bewitching, while in others hostile to the individual, always alarming, hiding within themselves, if not the destruction, then the corruption of the human soul, and at the same time the sweetness of this downfall.

However, with the same ease I'll begin copying out for you the same multitude of and equally brilliant motifs of the earth, forest, water, flowers, grass, steep slopes, hills, bees, sunsets (not urban), horizons and the free wind, clouds and foggy distances, clay hillsides and remote roads, dewy boundary paths and melancholy haystacks.

People assume that if there's a city in Blok's poetry, it has to be St. Petersburg, and if nature appears in his work, it refers without fail to Shakhmatovo. Despite the fairness of this view, I consider it a strained interpretation. Blok himself wrote:

> We remember it all—the Paris streets of hell
> And the Venetian coolness,
> The distant aroma of lemon groves,
> And the smoky masses of Cologne . . . [18]

To be sure, St. Petersburg and Shakhmatovo represent the two wings of Blok's poetry, but he soared on them easily and widely, at such heights that he could see farther than the two purely geographic points to which we sometimes want to limit him. There exists the attempt to force the landscape of Shakhmatovo even into the poem "On Kulikovo Field":

> The river spread out. It flows, grieves lazily
> And washes the banks.
> Above the barren clay of the yellow precipice
> Haystacks grieve in the steppe.
>
> . . . .
>
> And eternal battle! We dream of peace only
> Through blood and dust . . .
> The mare flies, flies through the steppe
> And tramples the feather-grass . . . [19]

It's true that in Shakhmatovo there is a river (the Lutosnya), and possibly there are meadows on the banks of this river and haystacks in these meadows, and it doesn't matter that one won't find any feather-grass in Klinsk District and that one can't, by any stretch of the imagination, call a horse of the Moscow region a "mare of the steppe." What does matter is that the landscape in the poem and the very image of Rus are so far removed from the Shakhmatovo lilies of the valley not only in appearance but also in spirit, that it would indeed be straining the interpretation to insist that the woody, shady, blackish Lutosnya served as the prototype for the Nepryadva, even though the poem was actually written at Shakhmatovo. As if one cannot live among woody hills and at the same time retain a generalized image of the Russian land in one's mind.

Similarly unfounded, it seems to me, is the attempt (and one encounters it on occasion) to identify the Shakhmatovo garden (which indeed boasted myriad nightingales) as the nightingale garden of Blok's poem of the same title.[20]

The harsh sun of southern France; white, burning-hot stones; stratified cliffs; and juxtaposed with all this—the deep-blue twilight of a shady garden behind a stone fence, in which may not be mentioned, but are conjectured and imagined, babbling fountains, while the streams along the little roads are even mentioned—even in the face of the obvious symbolism in the poem, all of these things are images and symbols of another order, from another world rather than from Blok's actual garden, which from its lindens and poplars imperceptibly turns into a dark fir forest and is virtually fenced off from the remaining world of clover, meadows, and soft coolness by a spinning wheel made of two poles, and in which, in the quiet evening dew, each sound from Osinki and also Gudino can hardly be heard; and these sounds include the whetting of a scythe, the jingling of the well's chain, and even the coughing of an old woman, as are mentioned in the poem "Autumn Day":

> I walk through the stubble-field slowly
> With you, my unassuming friend,
> And my soul is full of emotion,
> As if in a dark village church.[21]

Now this is definitely Shakhmatovo, and the church is undoubtedly the Tarakanovo Church, a photograph of which was sent to me not long ago.

The Blok scholars of St. Petersburg (now Leningrad, of course) seem to be in opposition to the Moscow school, if one can call it that, headed by the meticulous, astute, and tireless researcher Stanislav Lesnevsky. His two-volume study, *The Moscow Region in the Life of Alexander Blok*, will undoubtedly be enormously interesting, and we impatiently await its publication.[22]

However, we ourselves don't separate Blok into his component parts, even though we acknowledge that for the poet Shakhmatovo represented the Russian earthly font, so that it's possible that here, under the influence of the beauty of nature, there occurred in his soul that displacement, as a result of which (out from under a layer that gave way) the clear and abundant spring of poetry began to flow.

But let's rest assured that this was not at all the same Shakhmatovo that appears to us in Maria Andreevna Beketova's conscientious description. Even though she includes a chapter entitled "Walks at Shakhmatovo" in her *A Family Chronicle*, it still emerges as a small, closed world: the estate, house and garden, outbuildings, and Podsolnechnaya Road as a necessity, and the little neighboring villages as existing at a happy distance from the estate.

For Blok, Shakhmatovo least of all was limited to the estate. He "lived" in the house only in the narrow sense of the word: he ate, drank, wrote poetry and letters, planted trees and roses, mowed, used a hammer, sawed, and felled trees. The dwelling place of his soul was—if we can call it thus—"Greater Shakhmatovo" that is, Shakhmatovo with all its surrounding scenery, from the hamlet of Podsolnechnoe to Rogachyov, from Boblovo to Tarakanovo, from Runovo Rock to the Aladyisk Heights, and from horizon to horizon.

Maria Andreevna could live in a world of a single weeping birch and spreading mountain ash, of old linden trees and parterres of dog-rose bushes; Blok lived in the world of the Lutosnya, woodland swamps, roads and paths, hillsides and steep slopes, tall weeds, thickets of willow-herb, distant nocturnal lights located upriver, and the flashing glance of a peasant woman from under her printed scarf while she was walking along the road in the daytime.

After all, when abundant white fog would rise from the Lutosnya in the evening and disperse, enclosing in itself, like a lake, all the lowlands between the woody hills, and a reddish moon would float over this fog; when it would have been strange for Blok's aunts to find themselves outside the limits of the cozy house, moreover, the estate; it was precisely then that the young, strong, and handsome Blok, upon returning from a simple walk, and later from Boblovo, could find himself alone in fog-covered thickets in the forest.

> In the damp evening fog
> Only woods, and woods, and woods . . .
> In remote, damp weeds
> A light flashed—and disappeared . . .
> It flashed again in the fog
> And I thought I saw:
> A hut, a window, and geraniums
> So red on the sill . . .
> In the damp evening fog
> Toward the glowing red light,
> Toward the scarlet geraniums
> I guided my steed . . .[23]

97

Even in her dreams Maria Andreevna would not have imagined such a scene. To take a walk as far as Praslovo Forest by the boundary path for the oat field in a full summer dress under a brightly colored umbrella, to gather unhurriedly a bouquet of field flowers—these things were in the domain of the aunts. But to walk in noctural fog and tall weeds, in a damp forest, entrusting oneself only to the instincts of a horse in the midst of swampy places. . . .

For Blok the estate's boundaries were expanded initially with the help of his grandfather, Andrei Nikolaevich Beketov. An excellent botanist, he would take the boy along with him through forests and swamps, hills and streams. On these occasions they would gather flowers and plants, not for bouquets, but in order to learn about the world. Immediately would follow the Russian and Latin names of each plant and its membership in a species, family, and class. An element of play consisted in finding a plant that had not yet been discovered in those places near Moscow.

Whether yielding to the rules of the game or actually discovering rare species, Blok himself testifies as follows:

> For hours on end he and I would wander through meadows, swamps, and dense forests. Sometimes we would walk for dozens of versts after getting lost in the forest. For our botanical collection we would dig out grasses and cereals with their roots; upon doing this, he would name the plants and categorize them, teaching me the rudiments of botany, so that even today I remember many botanical terms. I remember how happy we were when we found a particular flower of the early *grushovka*,[24] a species unknown to Moscow flora, and the smallest low-growing fern; I still look for this fern on that same hill, but so far can't seem to find it— evidently it had seeded itself by accident and later died out.

To devote so many words to these childhood walks with his grandfather in a very short autobiography[25] is tantamount to attributing great significance to them. It is well known that a glance not interested in anything specific glides superficially over nature and her beauty, not really penetrating beyond her outward appearance to the depths, the insides. But with a concrete interest, even a trifling one (assembling a herbarium, collecting butterflies or birds' eggs, searching for medicinal herbs, catching fish), the superficial glance becomes a penetrating one, and a world thus far unknown to the individual is revealed. One can compare this to a simple fascination with the sea when a swimmer's glance glides over its surface, and with the amazing transformation of the sea when at that very second the same swimmer peers through the glass of his or her mask into an abyss illuminated by the sun, shimmering in its dark-blue color, then in its depths becoming gloomy—an abyss in which each marine plant, each little fish, each pebble on the bottom, when taken together, create a fantastic and enchanting landscape.

In circles that increasingly widened the distance from the house and garden, Blok became familiar with the surrounding fields and forests. The large number of hills enabled him to view the land from various and numerous projections, so that new scenes kept opening up to his enraptured soul.

There was a certain vantage point (on the hill opposite the village of Novaya?) from which a person could see at the same time twenty little white churches and bell towers scattered throughout the dark verdure of the hills and valleys. Can one imagine the hour right before vespers, when the bells would all chime together? Can one imagine them in a golden autumn? In the early emerald-green verdure of spring?

The flatness of the hills at various points is defined by the light. Some places are brightly illuminated, some are partly shaded, while others remain entirely in the shade. All this complicates the landscape, makes it as complex, agitating, and mighty as a symphony—almost like music (when one considers the participation of fleecy clouds, storm clouds, clearings in the sky, sword-shaped rays flashing through these clearings, and the wind disturbing the foliage). Or almost like Blok's poetry:

✦ ✦ ✦

I set out on a path with spacious views,
The wind bends the pliant bushes,
A broken stone lay down on the slopes,
Barren layers of yellow clay.

Autumn has cleared in the wet vales
Baring the graveyards of the earth,
But from afar glows the red color
Of dense mountain ashes in nearby towns

. . . .

Many of us—young, free, and graceful—
Die without having loved.
May your boundless distances shelter us!
Could we live and weep without you?[26]

✦ ✦ ✦

When in damp and mildewed foliage

99

A stand of mountain ash shines crimson-red—
When the executioner's bony hand
Drives the last nail into my palm—

When above the leaden ripple of the rivers
At a damp and gray elevation
Before the face of a motherland severe
I begin to sway on the cross. . . .[27]

Such was the Shakhmatovo of the poet Blok.

An interesting perception of Shakhmatovo, and of Blok in it, is that of another Russian poet and friend of the Bloks, Andrei Bely. In summer of 1904 he visited Shakhmatovo; I might mention, incidentally, that he apprehended this place, not without bias, but under the indisputable influence of Blok's poetry; he apprehended it—I fear to say—literarily:

> The mystical mood of Shakhmatovo's surroundings is such that here one feels a virtual struggle, an exclusiveness, a tension; one senses that here sunsets stand out in a different way against the jagged peaks of woody hills; one senses that even the forests themselves, full of swamps and swampy windows into which one could fall and perish irrevocably, are populated by all sorts of evil spirits ("swamp sorcerers"—and little demons). In the evenings the "Invisible Being" appears, but the first glow of sunrise with a bright ray reveals the dual nature of the woody swamp. I describe the style of Shakhmatovo's environs because they are reflected so clearly, precisely, and realistically in the work of A.A.[28] The landscapes in most of his poems (*Poems about the Beautiful Lady* and *Unexpected Happiness*)[29] belong to Shakhmatovo. . . .
>
> . . . I only remember that, while driving toward Shakhmatovo and noting the connection between its landscapes and those in the poems of A. A., A. S. Petrovsky and I began to experience a romantic frame of mind. . . .
>
> . . . In this frame of mind we drove right up to Shakhmatovo, whose estate, structures, and outbuildings emerge almost imperceptibly, as if from out of the forest, concealed by the trees. . . . Our *brichka*[30] entered the yard and we found ourselves at the gray, wooden porch of a small, one-story house with a second-story mezzanine comprised of two rooms, in which A. A. and I later lived.
>
> I recall that my impression of the rooms where we wound up was that they were cozy and bright. The layout of the rooms was predisposed to coziness; it was the layout of the small houses I knew and loved so well, where everything exuded the simplicity of the past culture and way of life of the nobility, and along with this a timelessness: one sensed that in everything within these walls, pure "walls," that is, conditional and temporal boundaries, there also existed boundary

paths into the "golden absence of roads" of a new time—nothing was specifically old, such as portraits of ancestors, furniture, etc., which create the stuffiness and dreariness of many country estates; but there also wasn't anything of the *raznochintsy*,[31] either—in everything there was an intellectual quality and sparkling cleanliness. . . .

. . . We came out onto the terrace and went into the garden, which was located on a hill with steep little roads that almost became narrow forest paths (a forest surrounded the estate), walked through the garden and came out in a field, where we saw A.A. and L.D.[32] returning from a walk. I remember that their image in relief became fixed in my mind: on a sunny day, among flowers, L.D. in a wide, fitted, pink housedress that especially suited her and carrying a large umbrella in her hands, young, rosy-complexioned, strong, her hair streaked with gold, and with one hand lifted to her eyes (evidently trying to discern who we were), reminded me of Flora, or a pink Atmosphera—there was something in her appearance from some lines in A.A.'s "Blossoming Dream" and "Golden Locks on Her Forehead . . . ," and from the poem "Night Is Falling, Believe."[33] A. A., who was walking beside her, reminded me of that fairy-tale tsarevich about whom tales prophesied: he was tall, well-built, broad-shouldered, suntanned, I think, without a hat, his health improved by country life, in boots, in a finely sewn, oversized Russian white shirt embroidered by his mother's hands (I think the design was that of white swans on a red border). "The Tsarevich with the Tsarevna"—this is what involuntarily burst forth in my soul. This is how I remember that sunny pair among wildflowers.

. . . In A.A. I sensed here once again (as I sensed more than once in different circumstances) not romanticism but a connection with the earth, with the Penates of local places. Right away it was evident that he had been raised in this field, garden, and forest, and that the natural landscape was only an extension of his rooms, that the Shakhmatovo fields and sunsets were the actual walls of his study, while the magnificent bushes of the bright-crimson dog rose with a golden core that I had never seen before, against which background this young and healthy couple now stands out, constitute the real stylistic frame of his fragrant lines. Into the rosy-golden air of this peaceful atmosphere to which I was privy even in Moscow, now intruded the heady scents of the Shakhmatovo flowers and the rays of the warm July sunlight—"singing and flushed, she stepped onto the porch"[34] was written by him here; I always thought that the lines originated here. . . .

. . . I looked out the window over the tops of trees near a corner of the garden where the land sloped downhill, at the horizon of an already pale-blue sky with slightly golden, ash-gray clouds—I saw the summer lightning flashes of "Golden feathers of clouds: the dance of tender night-school students."[35] In a word, the first day of our stay at Shakhmatovo passed as if it were a reading of *Poems About*

the Beautiful Lady, while the entire succession of days at Shakhmatovo constituted
a cycle of Blok's poems.

Yes, Andrei Bely's perception of Shakhmatovo, if we judge by these recollections, was a literary one, secondhand through Blok's poems. But Andrei
Bely perceived Blok himself—that is, his poetry, of course—one-sidedly, from
the side of his symbolic bell tower. Does the pathos of Blok's poetry really
exist in these "blossoming dreams," "night is falling," and "golden locks on her
forehead"? Bely's Blok emerges as the kind of poet who describes roses and
daydreams, the coziness of an estate, fragrant lines, the rosy-golden air of a
peaceful atmosphere, into which the heady scents of Shakhmatovo flowers
seem to intrude.

To be sure, it was still 1904. "Autumn Freedom," "Lifeless Old Age Wanders
Everywhere," "A Girl Was Singing in the Church Choir," and "In Shaggy and
Terrifying Paws"[36] had not been written yet; all this would be written one year
later, in the summer of 1905. Moreover, the entire cycle *Homeland* had not
been written, and "On Kulikovo Field" with the Nepryadva had not yet
become a part of Blok's poetry; perhaps all of this indicates that Andrei Bely
had not really been mistaken. Nowadays we perceive Blok in toto, as a manifestation, with his heights—his "ceiling," as aviators say—the entire breadth of
his work; but at that time he was just getting started and, even in approximate
terms, had not said what was most important to him.

But still, even then it was already possible, if not to see, then to sense that
Blok was not at all a bard of rosy-golden air, but that, on the contrary, he was
a poet of discomfort, of the wind whistling through bare branches, of heavy
approaching clouds, of autumn graveyards, clayey hillsides, bloody sunsets,
and the anxious cries of swans—that he, in short, was a poet and prophet of
impending doom.

At the same time, he was also a life-affirming poet, though by no means as
A. Bely saw him; rather, his love of life was vivid, active, energetic—a love of
life with an axe in its hand, with a scythe, riding a horse, a love of life with a
blinding smile, with a face turned toward the wind. "I hear a bell. It's spring in
the field. / You opened cheerful windows," "I get up on a foggy morning, / The
sun strikes me in the face, / Is it you, my desired friend, / Stepping up on the
porch to see me? / Wide-open gates are heavy! / The wind blew in through the
window! / Such lively songs / Haven't been sung for a long time!" "Having scattered across the horizon's sky, / the fiery-red cloud moves"; "This iron staff has
been lifted / Over our heads. And we . . . "; "She galloped through the wild
steppe / On a lathered horse"; "How long will you clank the chain? / Come and
dance with me!"[37] . . . Now where in these lines, one wonders, is the rosy-gold-
en atmosphere with the intoxicating scents?

And, in addition, what does Shakhmatovo signify in a cycle of poems by Alexander Blok? As we were taught in school, Shakhmatovo is nothing more than objective reality. Based on this reality, one person will write this kind of poetry, while another—a different kind. Moreover, we have an example for comparison: one can't invent one deliberately. Ekaterina Andreevna Beketova (Krasnova), as we know, wrote poems and even published them in a collection that was awarded the coveted prize of the Academy of Sciences. And so all the poems of Ekaterina Andreevna are imbued with Shakhmatovo.[38]

And thus, does one sense "the mystical mood of the environs" in her poems? Does one sense "a kind of struggle, exclusiveness, tension"? That "here sunsets stand out in a different way against the jagged peaks of woody hills," that "in the evenings the Invisible Being 'appears,' but the first glow of sunrise reveals. . . . " And so on? Of course not! These are the usual, charming poems of a cultured woman of the nineteenth century, a lady and a member of the intelligentsia, I would say. This means one of two things: either the mystical moods existed in Blok's soul and they painted the landscapes in his poems with special colors, illuminated them with a certain light; or these moods lived in Andrei Bely, who under their influence read Blok's poems in a special way, seeing in them what was not there.

It is also worthwhile to take a look at Ekaterina Andreevna's poems in order to see how the same strings under the fingers of a dilettante produce merely charming sounds, whereas under the mighty hand of an inspired and brilliant master these strings rumble and resound.

The best-known poem by Ekaterina Andreevna is already well known because Rachmaninov set it to music, and it exists today in the form of a romance entitled "The Lilac Bush." This lilac, it turns out, is located at Shakhmatovo:

> In the morning, at sunrise,
> Along the dewy grass
> I go to breathe fresh morning air,
>
> And in the fragrant shade,
> Where dense lilacs grow,
> I seek my happiness. . . .
>
> In life I am fated
> To find one kind of happiness,
> And this happiness in lilacs does live;
>
> On green branches

In fragrant clusters
My poor happiness blooms.[39]

It's charming, isn't it? She also wrote verses about the Shakhmatovo nightingales. Here is an excerpt:

In the evenings of a flowery spring
A nightingale flies to our yard,
Where the fragrance of lilacs
Combines with the cool of the night.

When the air is warm, fragrant, and clear,
Quietly open the window to the yard
And you'll hear the bird's voice so sweet
As it sings from sunset 'til dawn.

And you'll see in a clear sky
The new moon as it glows and burns,
And the apple tree in its fragrant attire
Stands hoary in the light. . . .[40]

This is the poetry of quiet, secluded estates. "Quietly open the gate . . . ," "The chrysanthemums already bloomed long ago in the garden . . . ," "It's autumn. Our poor garden is completely losing its leaves . . . ," "Looking at the ray of a purple sunset. . . ."[41] These are all poems of the same kind—some slightly better, some slightly worse than those of Ekaterina Andreevna Betekova:

Just yesterday the forest that had grown bare
Sadly said good-bye to me
While dropping a yellowed leaf
Until its joyful encounter with spring.

My path was all covered with leaves
From a rain that was silent and golden,
And the trees whispered quietly to me
Asking me to return to them.

Parting was so hard for us,
But from the sky and fields far away
So pleasant, so sad, and so thrilling
Suddenly came the call of the cranes. . . .[42]

I admit that I've somewhat taken advantage of the reader's attention, but after all—she's Blok's own aunt! She has the same genetic code; the relay-race flame of poetic gifts passed through this stage out of the darkness of previous generations, as a flame passes along Bickford's fuse, and with a blinding explosion it arrived and illuminated not only the environs of Shakhmatovo but all the borders of Russia.

Incidentally, in all fairness I should add that one of Ekaterina Andreevna's poems (I leafed through her entire collection, a bibliographic rarity not in danger of being reprinted in the foreseeable future) is constructed according to a genuine poetic idea, so that, if one didn't know in advance, one might take it for an unknown poem miraculously discovered in the archives of, say, Tyutchev. I think that it would completely fit in with his work:

> Against the pale gold of sunset
> The jagged forest's wall grew dark.
> And, filled with a dark blue haze
> That merged with the cupola of the skies,
>
> A sea of fields already ripened
> Stretched in every direction,
> And was agitated in the expanses
> In the glow of fading rays.
>
> The sunset was gone. . . . But the undying light
> Already started to shine on the earth
> And illuminated the evening dusk
> That lay imprinted on the fields.
>
> And from above the sky looked down,
> Dressed in its evening mantle,
> As the waves of golden grain
> Brought light to the earthly gloom.[42]

As God is my witness, I copied out this poem for the sake of fairness and to the detriment of the material I have presented here. After all, what I need now is a sharper contrast between the poems of Ekaterina Andreevna and her nephew, a more advantageous one, as this part of my essay is constructed specifically according to contrast. But let us hope that the reader hasn't yet forgotten Ekaterina Andreevna's nightingales, or her lilacs, or the basic tone and level of her poetry.

And now what follows represents the same source of inspiration, seemingly the same strings, even the same clover, but the sound is different:

> I sank into a sea of clover,
> Surrounded by the stories of bees,
> But my childish heart was discovered
> By the north wind that beckoned to me. . . .

The entire enigma of this poetry (and its entire sense and meaning) lies in the fact that the very same words and the very same subject matter are suddenly rearranged, restructured on different levels, and transformed into a different quality. Similarly, the same bricks, after being rearranged, are transformed into a gloomy tower on a cliff or the epitome of a Gothic cathedral in place of an idyllic little house surrounded by greenery.

> Your innermost melodies
> Hold the fateful message of doom,
> The damnation of sacred requests
> The profanation of happiness.
>
> But it's such an enticing force
> That I'm ready to repeat the rumor
> That you brought angels to earth
> By tempting them with your beauty.
> . . . .
>
> I wanted us to be enemies—
> Why, then, did you give me
> A meadow with flowers and skies filled with stars—
> Your beauty's summary damnation?[44]

Well, all right. I admit that here the generalized moment is a little too strong and the entire poem is written, on the whole, about an abstract topic, about the Muse. Let's take a poem concretely linked with Shakhmatovo and consider whether it's possible to measure the distance from it to the usual lines about the scenery that are populated with carnations, wild strawberries, and multicolored lights:

> Lifeless silence wanders all around,
> A path has drowned in verdure,
> I'm upstairs sawing a semicircle—

I'm sawing a dormer-window.

I smell the distance—and drops of pitch
Emerge in the veins of the pine wood,
The shrieks of the saw intrude
And golden sawdust flies all around.

Here's the last, whistling split—
And the board flies into the unknown . . .
In the sharp scent of melting pitch
The outdoors opens up before me. . . .[45]

Blok at first was considered a symbolist poet only out of misunderstanding; only the symbolists themselves, with their listless and, in general—I don't hesitate to say—dreary poetics would have liked to consider him one of them. Blok was really just a master who was able to build words into musical (as they could sing only in Blok's work) lines, and these lines into similarly musical (though iron-clad in their organization and purposefulness) stanzas.

I don't remember who it was who, having spent some time in Blok's apartment and study, and expecting to see there some sort of bohemian, symbolist chaos, or at least disorderliness, was struck by the exemplary—to the point of pedantry—order both on his desk and everywhere else, by the scrupulous cleanliness and the almost monastic, ascetic austerity.

Blok was remarkably successful at creating the lead-ins to his poems, the first lines, which talent, incidentally, was imitated by his first student, Sergei Esenin, whose link with Blok's poetry has not been studied—and it is much deeper than one can surmise from a superficial glance. One can walk about all day with the introductory lines of one or another poem by Blok—repeating them, reveling in them, and deriving real joy from them:

✦ ✦ ✦

In these white nights, my cruel,
Eternal knock at the gates: Come out!

✦ ✦ ✦

You walked away, and in the desert
I nestled against the hot sand

✦ ✦ ✦

I move from torture to torture
Of fire's broad band.[46]

✦ ✦ ✦

I am a shivering creature.
Dreams stagnate, revealed by rays of light.

✦ ✦ ✦

Why have you bowed your head in confusion?
Look at me as you did before.

✦ ✦ ✦

No one would call me insane;
My bow is low, my countenance—severe.

✦ ✦ ✦

At the door I met the unfaithful one:
She dropped her scarf—and was alone.

✦ ✦ ✦

Everything transient, everything fragile
You buried for all time.

✦ ✦ ✦

Oh, spring without end, without limit—
Without end, without limit my dream![47]

Let readers look through Blok's poems from this perspective. It stands to reason that after a cursory perusal not everyone, perhaps, will be caught up in the
power of the music, not everyone will be swept away by the bright wave; but
even so, after several days have passed, suddenly and unexpectedly, as if from
nowhere at all, all of a sudden, in the course of bustling daily concerns, in the
soul will resound:

The sound approaches. And, submitting to the melancholy sound,
The soul grows younger.[48]

But we've gotten carried away. It is not Blok's poetry, not his creative work itself that concerns us now, but first and foremost Shakhmatovo.

At Shakhmatovo Blok wrote about three hundred poems, not counting his letters, diaries, notes in notebooks, and articles. But it would be simplistic and even unprofessional to separate the poet's poems into those that, in their essence, are associated with Shakhmatovo and those that are not. Only those completely ignorant people who are very far removed from the literary trade (as writers fashionably call it these days, but nevertheless—it's art, art!) are inclined to think that if a writer travels to Ryazan and settles there for the summer in some nearby village, it means that, right away, without fail, he or she will begin to write about Ryazan. But meanwhile the writer is writing about last year's impressions of a trip to Siberia; or, in general, about the Cologne Cathedral. For example, Blok's poem "To the Muse," from which I have cited several stanzas, in its spirit belongs to Shakhmatovo (a meadow with flowers); however, it dates to the end of December 1912, when Blok could not have been at Shakhmatovo. I have already mentioned that, although the poem "On Kulikovo Field" was written at Shakhmatovo, it is not at all imbued with the landscape of Shakhmatovo. In its entirety it evokes steppes, feather-grass, wormwood, and *The Lay of the Host of Igor*.[49] Is the poem "A Girl Was Singing in the Church Choir" set in the Tarakanovo church? It is dated "August 1905." More than likely, it is set in the Tarakanovo church. In my presence Stanislav Lesnevsky persistently questioned the local residents concerning whether there had been in the Tarakanovo church, above the iconostasis, a wooden sculpture of a little angel, a little cherub, having in mind the last lines of the poem (" . . . and only above, near the Royal Doors, privy to their secrets— / a child was crying, because no one would return"). But isn't it possible that these lines were written from memories of an impression that had been experienced? Or from the confluence of two impressions: an old one and a fresh one? To be sure, Blok quite often writes realistically in his poems; quite often his poems represent a poetic diary: uninterrupted, detailed, sometimes two or three poems per day. But still, the poet recorded, not so much the external event as the movement of the soul, albeit engendered by an external event; in reading the poem, however, one cannot always identify and interpret the external event. It is said that "A Girl Was Singing in the Church Choir" was written in the days when Blok was grief-stricken over the unfortunate news of the death of Russian sailors in the Tsushima Strait. What does one make of this? In its breadth and profundity, and its generalizing moment, the poem goes far beyond the frame of a concrete event, even if it is a major national tragedy.

Blok started writing the poem *Retribution* on a rock near the hamlet of Runovo (the hamlet no longer exists, but the rock is still there; it stands on high ground, one can see a great distance from it, and Blok liked to sit on it). And when, in the introduction to the poem, Blok overwhelms us with his mighty iambs—

> The final judgment is not yours,
> It's not your place to seal my lips!
> And let the church be dark and empty,
> The pastor sleeping; 'fore the Mass
>
> I'll walk the dewy bound'ry path,
> The rusty key the lock will open
> And in dawn's crimson vestibule
> My Mass I'll start to celebrate—[50]

—when we read these lines, we understand that Blok, in taking walks in the morning, evidently more than once walked along the dewy boundary path from Shakhmatovo to the church at Tarakanovo, although he didn't go inside, because how could he have entered a locked church? And if someone had let him inside, then he would not have been alone in it. But in his thoughts he could enter it anytime he wanted; in any case, he entered it in the poem *Retribution*.

Incidentally, in "Confession of a Heathen"[51] Blok testifies: "And I also used to go to church at one time. Of course, I chose a time when the church would be empty. . . . In an empty church I sometimes succeeded in finding what I sought in vain in the world."

Apropos here is an example of how the same sensation, the same thought, is expressed in prose and in poetry.

The Tarakanovo church (you see, I keep writing about it in order to argue for its possible renovation) also entered Blok's biography in the form of a more serious event, one of the major events in the poet's life. And even Shakhmatovo itself, regardless of the magnitude of its significance for the formation of Blok's soul and manner of his thinking, would have lost a large part of its memorial fascination if the hamlet of Boblovo, where Dmitry Ivanovich Mendeleev lived, had not stood seven versts from it on a high hill that dominated the surrounding landscape (as military topographers would put it).

The great scientist bought this estate in 1865; people say that he bought it because of the magnificent views that opened up from the hill. He only came to take a look at it, but as soon as he found himself standing on the hill facing the Russian land that spread out before him, with its hills, valleys, forests, multitudes of simultaneously visible little villages and churches, and with its fluffy clouds, he didn't want to leave this place. Somewhere over there, far away below, seven versts away (entirely through forests), invisible from this place was the little estate of Shakhmatovo, which only nine years later Mendeleev would advise his friend, the professor and botanist Beketov, to acquire.

At Boblovo, Mendeleev had a spacious house and a well-equipped laboratory, in which Dmitry Ivanovich conducted experiments in meteorology, agrochemistry, and pure chemistry. And indeed, for the scientist all the fields at Boblovo represented a distinctive laboratory, if one has in mind the fields attached to the estate and not those owned by the peasants of the hamlet of Boblovo, which was located not far from the Mendeleevs' park and house, but still on the same high hill.

Meanwhile time was passing. At neighboring Shakhmatovo, surrounded by loving and educated aunts, and likewise by cousins his own age (they played American cowboys and Indians), the handsome light-brown-haired youth was growing older. There was running around in the garden, short walks with his aunts and long walks with his botanist grandfather, and later on, solitary walks on foot and rides on horseback. In circles that kept getting farther from the house and garden he became acquainted with the surrounding land, with its swamps, ravines, forest paths, expansive meadows, streams, and glades. If Aksakov had frequented these places, he would always have traversed them with a rifle and fishing poles; he would have learned where all the deep pools in the Lutosnya were; and he would have known where perch take bait, where roach-fish could be found, where hazel-grouse would nest, where heath-cocks performed their courtship rituals, and where woodcocks would call. Turgenev as a hunter and Chekhov as a fly fisherman would have appreciated these places. But it's difficult, even impossible, to imagine Blok with a rifle or a fishing pole. His spirit was uneasy, restless, prophetic, anticipatory (despite the visible well-being and blossoming) of approaching cataclysms and even seemingly awaiting them with impatience: "I've worn out the sad, nocturnal path / That leads to the country graveyard."[52] This means that unknown even to the inhabitants of the Shakhmatovo house were regular nocturnal walks to some nearby village, to a graveyard, and brief sojourns there in solitude and silence. The proximity of the graves and crosses of the graveyard church contributed to the mood of these nocturnal walks. Let us try to imagine the same Turgenev, Chekhov, Nekrasov, or Fet at this place—it doesn't work, one cannot visualize it. But one can see Blok with his arms folded in the shade of a graveyard church, as in a painting.

> I'm on a ledge. Above me—a grave
> Of dark granite. Below me—

A path that appears white in the dusk,
And anyone who looked at me from below

Would be startled: I'm so still,
In a broad-brimmed hat, among graves in the night,
With my arms crossed, handsome and in love with the world.[53]

This is from the poem "Above the Lake," which is set somewhere in Finland. But didn't he stand in precisely the same way above the Shakhmatovo valley, which, moreover, was famous for its nocturnal fogs that developed no worse than those of other lakes?

Impressions from these walks fill the lines of his poems:

✦ ✦ ✦

In the wild grove near the ravine
Is a green hill where it's always shady.

✦ ✦ ✦

I walked into bliss. The path shone
With the red glow of evening dew.

✦ ✦ ✦

The white steed barely treads with tired feet
Where boundless ripples have formed.

✦ ✦ ✦

The quiet of fading grain,
This radiant time in the world.

✦ ✦ ✦

The sky is glowing. The silent night is lifeless,
Masses of trees in the woods surround me.

✦ ✦ ✦

I ascended all the heights,

And looked at different skies,
My torch was the eye of an owl
And the heavenly dew of the dawn.[54]

◆　◆　◆

I look for lights on the way
To the black limits of your mind.
The huge moon has turned red
Among dark, muddy creeks. . . .

His double floats above the forest
And soon will turn to gold.
And then—the domain of the swamp demon,
Water demon, forest demon. . . .

The path continues, the moon is higher
And stars fade in the silvery sky.
And softly shine the roofs
In the darkling village on the hill.

I walk while dewdrops shiver,
Thinking silvery thoughts of you,
Just of you unplaiting your braids
For your secret lover in the cottage.[55]

◆　◆　◆

In the damp evening fog
Toward the glowing red light,
Toward the scarlet geraniums
I guided my steed . . .[56]

I don't know whether what is in these poems about a secret love resulting in unplaited braids in a cottage and a little house with geraniums in a damp, nocturnal forest was only imagined or already real (and why couldn't it be?), but the circles of walks at Shakhmatovo keep expanding until one day they lead the young poet—a handsome and romantically inclined youth, a well-built rider, a kind of prince and knight of the Shakhmatovo hills—to a high mountain in the direction where the Shakhmatovo sun would usually set, and where the evening sunlight would usually shine above a dark, jagged forest. In his

113

prose Blok describes this portentous moment for us:

> We descended to the bottom of the ravine. Gray [Blok's horse—V.N.] jumped across a rivulet that rushed through stones over yellow sand and jumped up onto a steep slope that ran along the other side; here lay a road along which I had not ridden previously. Gray didn't know which way to turn, either—to the left or to the right, and so he stopped. I let him walk in the direction that I imagined would lead us farther from home. . . .
>
> . . . On this road I immediately felt something that I loved and had forgotten, and began to imagine how tall the grain would be here in summer, the yellowish-blue carpets of cow-wheat and pink clouds of willow-herb. . . . Already I was entirely in the power of this new place. . . . I noticed that what had seemed to me to be a grove was an overgrown park, evidently part of an estate. I wanted to ride around it, and so I trotted along the ravine of clipped firs.
>
> Suddenly, to the right of the road behind several small logs thrown across a ditch, a little path appeared that led to the mountain through the tall trunks of firs and birches. I set out on it and, having reached its highest point, found myself facing new, massive expanses that opened up new plains, new villages, and churches before me.
>
> The park ended abruptly, and there began a series of structures not owned by peasants and a large yard of fruit trees all in bloom. Among the apple, cherry, and plum trees stood some beehives; the ravine was not high and was enclosed by old boards that were torn away in places. Quiet reigned here: not a sound reached this place from either the village or the estate.
>
> Suddenly an unexpected wind came up and scattered the colors of apple and cherry trees all around. Beyond the snowstorm of white petals that had flown onto the road I saw, sitting on a bench, a statuesque young woman in a pink dress, her hair in a heavy golden braid. Evidently she had been startled by the unexpected sound of my horse's hooves, for she quickly stood up and the color rushed to her cheeks; she ran into the depths of the garden, leaving me to watch her pink dress as it flashed beyond the snowstorm of petals.

Everything here has been somewhat romanticized. The description of the park, for example. Nowadays the park is in fact overgrown, but Mendeleev was a good caretaker and maintained his household in an orderly way. He had enough time for this, for his scientific work, for an ascent from Klin in a hot-air balloon in order to observe a solar eclipse (he landed on the territory of Saltykov-Shchedrin in Spas-Ugol of Tver Province), and for conducting agricultural experiments.

Couldn't the young horseman in all likelihood presume that he was located somewhere near Boblovo rather than guessing: Where had his wandering led him? In what kind of neglected estate had he wound up? After all, heaven knows that seven versts do not constitute such a great distance; and a friend of Beketov's grandfather[57] lived there, at Boblovo; and one could see the high mountain from Shakhmatovo, and there on occasion they would discuss Boblovo; and when Lyubochka Mendeleeva and Sashura Blok[58] were still children they would stroll together in the university garden in St. Petersburg under the supervision of their nannies. Whenever Mendeleev met Beketov he would ask: "Well, how's your prince doing? You know, our princess. . . . "

But the encounter was beautiful and romantic. Almost as if a presaging spirit had flown through the quiet garden, stirred up a snowstorm of petals, and instantly, as if materializing out of this snowstorm, there appeared a young woman in a pink dress with her hair in a golden braid.

Mendeleev's house was a house like any other, and Lyuba was like any other Lyuba—a healthy, rosy-complexioned, light-brown-haired girl. But now everything assumed a different coloration, a different illumination: a fantastic, jagged forest on the mountain, a high tower [the house—V.N.], a beautiful woman, Ophelia. . . .

In general, one must note that the German part of Blok's blood, from the Mecklenburgs, transmitted to the distant descendant vague memories of chivalry, a certain indelible watermark that became discernible and evident upon being held up to the light of poetry.

On the other hand, the heredity of a Russian nobleman (and, you know, initially in princely times, members of the nobility were warriors and bodyguards, and they received allotments of land specifically for their military service and thus became patrimonial landowners) made its voice heard through the Dark Ages. Gradually, the motifs of European chivalry of the Middle Ages—motifs of battles, the sword and shield—assume an increasingly Russian (once again I'll repeat the word "Lay-of-the-Host-of-Igor-like"[59]) coloration, until the heroic cycle *Homeland*[60] and the verses of "On Kulikovo Field" would resound like an organ. Here is this poetic evolution:

✦ ✦ ✦

I'm just a knight and poet,
Descendant of a northern skald.

✦ ✦ ✦

On this sorrowful earth I forgot
About valor, heroism, and glory. . . .

✦ ✦ ✦

A constant rustle surrounds the castle,
Transparent water stands in the moat.

✦ ✦ ✦

Here's the sword. It—was. But it's not needed.
Who made my hand grow weak?

✦ ✦ ✦

I died. I fell from a wound,
And my friends covered me with a shield.

✦ ✦ ✦

Few of us are left. We wear smoky cloaks,
Sparks fly and the chain mail shines.

✦ ✦ ✦

Will the heavy armor creak . . .

✦ ✦ ✦

Battle makes my heart grow happy,
I feel the freshness of military bliss. . . .

✦ ✦ ✦

Dearest knight, with snow-white blood
I was true to you.

✦ ✦ ✦

You as well can be entrancing,
You, the dark knight, you.

◆ ◆ ◆

I race away in freedom's air,
Weary from the heat of battle.

◆ ◆ ◆

Oh, love! You're stricter than fate!
More compelling than ancestors' ancient laws,
Sweeter than the sound of war's horn.

◆ ◆ ◆

Time to resume the previous battle,
Spirit—arise; flesh—fall asleep!

◆ ◆ ◆

I'm a sword sharpened on both sides.
A green stone flares up in my shield.

◆ ◆ ◆

He called me to the battle on a plain. . . .

◆ ◆ ◆

A cleansing wind is blowing
From the blue of the heavens.
The son throws down the fatal sword
And takes the helmet off his head.

◆ ◆ ◆

Yes, I'll meet you late at night,
Give you my hand in greeting,
You, who brought here from the fray
On a spear's point—spring.

✦ ✦ ✦

Once again in garlands and dewdrops
The dream starts to sing,
The gold of a shield
Shines again on the slope.

✦ ✦ ✦

A massive moon in the grass
With the reddening shield of the hero . . .

✦ ✦ ✦

Beyond the hill the sturdy armor stopped clanging
And the spear was lost in the gloom.
And my shield does not gleam—so golden and leathery—
This was all that I called my own.

✦ ✦ ✦

The son is guarded by the cross.
The son leaves his father's house.

✦ ✦ ✦

The tsarevna's song of spring—
So melodious, so inspired,
And I said: Beware, tsarevna,
For you shall weep for me.
But she placed her hands on my shoulders,
And I heard: No, forgive me.
Take your sword. Plan to fight,
I'll protect you on your journey.

✦ ✦ ✦

You and I stopped near the steppe at midnight:
We won't return and won't look back.
Beyond Nepryadva swans were calling,
And again and again they call. . . .

On the way—a flammable white stone.
Beyond the river—the dreadful horde.
Never again will the bright banner
Flutter over our forces.

And, bending his head to the earth,
My friend said: "Sharpen your sword,
So we won't fight the Tatars in vain,
We'll give our lives for the holy cause!"
I'm not the first soldier, or the last,
My homeland will suffer many years yet.
Oh, pray for me at early mass
Beloved friend, blessed wife!

✦ ✦ ✦

Once again above Kulikovo Field . . .

✦ ✦ ✦

And in the morning, when as a black cloud
The horde began approaching,
Your heavenly face was in my shield—
To shine forevermore![61]

In this manner certain, one might say abstract, castles, queens, swords and shields, manes and horns, helmets and battles of the Middle Ages gradually became infused with the deeply felt weight of Russian patriotism, while the almost equally abstract and symbolistic adoration of the Beautiful Lady assumed its own flesh and blood. All of this took place in the poet's soul, but it also took place in the woody hills between Shakhmatovo and Boblovo.

Afterward there were the daily trips by Blok to Boblovo, and Lyubov Dmitrievna also would come to Shakhmatovo, and in addition there was the family theater, at which they read poetry about Ophelia. Blok himself played Hamlet. The theater didn't even take place in the house, because for the performances they adapted the barn, which had shingles for its roofing material, and was cool, spacious, and clean, and contained hay.

They invited the peasants, too, and they always came: the men, women, girls, and kids, up to two hundred people. But . . . "the audience reacted to the performance in a way that was beyond strange. I'm talking about the peasants. During all of the emotional scenes in both *Hamlet* and *Woe From Wit* they would

laugh loudly, sometimes drowning out what was taking place on the stage"—
this is how Maria Andreevna Beketova remembers the performances.

She remarks about the actors with leading roles:

> They both recited the poems beautifully and played their roles with dignity, but
> on the whole they declaimed more than they acted. . . . Ophelia was dressed in a
> white dress with a square décolleté and light purple trimming. . . . In the scene of
> her insanity her slightly curled, loose hair had flowers entwined in it, and it cov-
> ered her lower than her knees. In her hands Ophelia held a whole sheaf of rose
> mallows, convolvuli, and hops, interspersed with other field flowers. . . . Hamlet
> was dressed in the traditional black, with a cloak and a black beret. A sword hung
> at his side.

All of this was probably beautiful and dignified, as Maria Andreevna
emphasizes, but of course very far from the comprehension of the Boblovo
peasants. And, you know, did they understand at all what was being said in the
emotional monologues?

But regardless of how things were, the theater existed, and, incidentally, not
only in the barn on a stage, but also in life itself. In having as a backdrop, now
the streets and cathedrals of St. Petersburg, now the hills and expanses at
Shakhmatovo, the action of another drama developed logically; its external
outline must have been more accessible, at least to the peasants, while at the
same time its secret essence and spiritual fulfillment were inaccessible even to
those closest to the action. It is possible that even the hero and heroine could
not give themselves a final reckoning concerning the events that had taken
place. The events had their own logic, and the actors submitted to it.

> The wedding was scheduled for eleven o'clock in the morning. As luck would
> have it, it was a rainy day and became clear only by evening. We all got up early
> in the morning and dressed for a festive occasion. The bouquet that had been
> ordered from Moscow for the bride didn't arrive in time. We had to put it togeth-
> er at home. Sasha and Mother picked some large, pink asters in the flower gar-
> den. The chauffeur, Seryozha Solovyov, ceremonially drove the bouquet to
> Boblovo in a troika of horses hired from Klin that had been prepared for the bride
> and groom. The troika was beautiful, tall, and light gray, with its shaft-bow dec-
> orated with ribbons. The coachman was young and elegantly dressed.
>
> Mother and stepfather blessed Sasha with the icon of the Savior. Aunt Sonya
> blessed him, too.
>
> The wedding ceremony took place in the ancient church of the hamlet of
> Tarakanovo. This was not a parish church of recent origin, but an ancient and
> noble one that had been built back in Catherine's[62] time. . . .

We all arrived at the church early and waited for the bride for a rather long time. Sasha was dressed in a student's frock coat, serious, focused, and ceremonial.

For this day they managed to hire some very respectable singers from the large village of Rogachyov. The rain had stopped and, while standing in the church at a side window, we could see all of the wedding guests as they arrived. They were all relatives of the Mendeleevs who lived right nearby. They all had brisk and fresh horses, whose shaft-bows were decorated with oak branches. The church was packed. And, finally, there appeared the troika carrying the bride, her father, her sister Marya Dmitrievna, and the little boy who carried the icon. She entered the church holding the arm of Dmitry Ivanovich, who had put on his medals for the occasion. He was extremely agitated. The singers began to sing "Go, my innocent one . . . " Yes, in truth, she was—an "innocent one.". . .

She was not married in the traditional silk, which would not have been suitable for a country environment, but wore a snow-white batiste dress that was fancy and had a very long train. Her dress was adorned with orange blossoms and she wore a bridal veil. It was impossible to look at this enchanting young couple without agitation.

. . . Dmitry Ivanovich and Alexandra Andreevna cried the entire time from tender emotion and in recognition of the significance of what was taking place.

. . . Upon leaving the church they were met by peasant men who brought them bread and salt, and white geese. After the wedding ceremony they rode off to Boblovo on their festive troika. As they were entering the house, the old nanny threw hops at them. . . . And meanwhile, in the yard was gathered a whole crowd of women dressed in their finery, who in their singing praised the groom, bride, and guests. Refreshments and money were sent out to them. When the champagne had been poured for everyone, Sergei Mikhalych Solovyov proposed a toast to the health of the young couple. . . .

The wedding took place in 1903; the last time Blok stayed at Shakhmatovo was in 1916. This means that thirteen of Blok's most mature, conscientious, and creative years were also years during which he spent time at Shakhmatovo. He no longer needed horseback rides to Boblovo with dreams and thoughts of a beautiful young woman and bride. The Beautiful Lady was now living with him in the same wing, which they redecorated in their own taste. The instinct to build a nest is not peculiar to birds alone: it inevitably intensifies when a change in life, such as marriage, approaches. Even previously, Blok had occupied himself at Shakhmatovo with activities that would improve the household; now this activity was especially natural for him. In his notebook we read:

Our wing of the house. Wild grapes. I need to cover the wall of the granary with meadow-sweet or philadelphus. Dig a little road. Cut down the linden tree. The bird-cherry trees. Young plant growth. Two flower gardens. Tobacco. Verbenas. Lilies. The philadelphus and lilacs on bare hillocks. Sakhalin buckwheat— on the back wall of the fence, up to the hazelnut tree. Mallows along the entire fence (seeds), plant Provence roses in the empty places. On the back wall— Sakhalin buckwheat. Birch trees. Silvery poplars.

It was just as typical for Blok to be in the garden with an axe or a spade in his hands as it was for him to be sitting at his writing table over a piece of paper. But he liked the axe better than the spade or saw. Cutting down trees and shrubs was his passion; once he chopped down an entire parterre of hundred-year-old lilacs. Lyuba gasped and was horrified. But it was all right. There would be more room, more air. And concerning his walks (when he and Lyuba lived in the wing), they became solitary, to distant places; he was especially agitated by the environs of the Rogachyov Highway. It is true that his wanderings had always been solitary (just a horse and rider) in the past, too, but in his thoughts he had imagined a rendezvous awaiting him and a tower on a high mountain; now his thoughts ranged far and wide. Freedom. Autumn freedom. This is the title of one of Blok's best poems, "Autumn Freedom." "I set out on a path open to views, / The wind bends the pliant bushes. . . . "

Manifestations of lofty poetry are sometimes determined largely by external, accidental, and ordinary circumstances. During Lermontov's childhood years an oak grove, or oak forest, happened to be located not far from Tarkhany. The youth liked to ride there on horseback and would spend entire days there under the broad, somewhat damp oak canopy that was filled with green light and green coolness. And hence at times one encounters an oak tree in Lermontov's verse. "The oak leaf has torn away from its native branch . . ."; "Let the dark oak above me always be green, arching, and murmuring."[63]

In Esenin one finds a similar predilection for the birch tree. And in reality, near Konstantinovo there grew (yes, I think that even today it's still intact, only they built a pigsty there) a splendid birch grove.

In Blok one finds grasses, steep slopes covered with dense forests, woody swamps, and hillsides, but especially fog. The Lutosnya River itself is not visible until one walks right up to its bank. It flows between banks covered with forests and alder thickets. But if there is a slight wind (it goes without saying—a warm, summer wind), white strands of fog will appear at the bottom of the valley. They pass through the trees, become entangled in the grasses, and accumulate. And immediately a sparkling white river will weave through the black forest, duplicating (but more broadly and in a more eroding manner) all the bends of the channel. The fog grows larger and more dense. The river of fog turns into a lake of fog. The fog rises almost to the midpoint of the hills, willfully and fantastically altering the entire landscape. At this time if the distant hill of Boblovo with its jagged forest and high "tower" is visible, then it is only above the fog, suspended in the air, floating, hovering, while behind it is the dawn. The magnificent fogs of Shakhmatovo!

However, a predilection for an oak, birch, or grasses with fogs is nevertheless insignificant in comparison to the most important thing that these native places gave the Russian poets. This main thing was a feeling for their homeland. Childhood impressions are the most vivid and most enduring: the foundation of one's future spiritual life, a stock of gold. In childhood the seeds are sown. Not all of them will sprout, not all will blossom. One won't notice them later in one's daily life, but they are there. The biography of a human soul is the gradual growth of seeds sown in childhood. Some of them grow into bright and clean flowers, and some into ears of grain, while others become vicious thistles. The life that follows is complex and multifarious. It consists of millions of actions determined by many character traits that, in turn, are formed by this personality. But if some kind of fantastic mind could trace and locate the connection between actions, it would find that every character trait of the adult person, every quality of his or her soul and, perhaps, even every action of his or hers, was sown in childhood and contained its own embryo, its own little seed from that time.

A feeling for his homeland, the perception of Russia as his homeland and the complete spiritual confluence with her constituted this kind of bright little flower that slowly, over the course of decades, blossomed and spread in Blok's soul. Several pages above we saw from examples of lines of poetry (this is why I copied out so many of them) how this occurred. "Oh, my Rus! My wife! Painfully clear / To us is our long journey!"[64] No one, either before or after Blok, ever referred to his homeland as a "wife" rather than a "mother," and it didn't sound blasphemous or artificial because of what he had endured, suffered through, and, if you will, cultivated in himself. Later, from this same background, came his "Scythians."

And the little seed came from Shakhmatovo.

III

Blok loved Shakhmatovo with all his heart. His answer to a question on a form, "Name a place where you would like to live"—"Shakhmatovo"—became a common passage in articles about him. And the fact that once (in 1910) the poet (for the first and last time) prepared to spend the winter at Shakhmatovo but ran away in late fall because of the "melancholy," doesn't mean anything.

Consider the following lines from Blok's letters to his mother:

> . . . we're having a snowstorm. There's already a lot of snow in the forest. Fyodor
> did some work on the second half of the pond, and they've sawed up the poplar
> tree. In the well [yet another attempt to dig out the well at Shakhmatovo, unsuc-
> cessful, just like all the previous ones.—V.S.] there are small stones in the damp
> sand, similar to those on a river bottom.
>
> The house is enclosed securely and cozily with screens and shutters. Without
> them life would be very unpleasant. We have completely moved into the annex,
> and dine in the small room. It's very warm. As soon as you left, the old house
> became huge and empty. Afanasy and I have installed locks everywhere, fixed
> the windows, and immediately a snowstorm came up. . . . Today we hiked on
> Mount Malinov and along the Praslovo. In winter the old forest reminds one of
> Heine. We bought *valenki*[65] for ourselves, and soon we'll have to buy skis, too. I
> haven't finished my poems yet. . . .

This letter was written on October 18, 1910, that is, at the beginning of
November, according to the New Style calendar. All summer they had made
repairs on the house and had built an annex onto it. Blok directed the work
himself, and it's entirely clear why he wanted to spend the winter in the newly
repaired and renovated house in which he had invested so much effort and
energy. It was tempting, interesting, and productive. However, as early as
October 22 he writes:

> . . . for two days we've had strong winds, and the house shook. This evening it
> almost became a hurricane, then the snowstorm came up, and by morning we
> were already walking in deep, quiet snow. Until now the outdoors have been
> unpleasant and irritating, but the snow has decorated everything. Now in the
> evening a thaw has begun. Snow is dripping from the roof and from branches; we
> built a snowman, and he kneels on his knees and prays, but very likely nothing
> will remain of him tomorrow.
>
> However, to spend the winter here would be impossible—mind-numbing bore-
> dom. Even the peasant men share this opinion. We go to bed early. Through-
> out this time I've recopied half of a collection of poems and have written a pile of
> letters. . . .

I shall repeat that these sentences in no way testify to Blok's dislike of his
Shakhmatovo. In general, it's difficult to spend the winter in the total isolation
of the countryside (and here it's not even the countryside, but the isolation of
a forest). In order to do this, one needs a particular psychological stability, but
mainly one needs to be accustomed to it. Throughout his life Blok was used

to spending the winter in St. Petersburg. The theater, literary evenings, constant interaction with friends, acquaintances, new acquaintances, women, wine, restaurants, city lights, cabdrivers and cabs, and, quite simply, St. Petersburg itself, with its atmosphere and mood, its myriad influences on the mind and heart. . . . An insufficiency of information, as we would put it today, spiritual, psychological, sensory, and so on, would of course be distressing for Blok during the long winter months, even if he spent them with Lyuba.

Similarly meaningless are the sentences Blok wrote about Shakhmatovo already after the Revolution, when Shakhmatovo no longer existed. As if he were telling his friends: "That's the way it goes," and, in addition, "A poet shouldn't have anything."

One needs to know of this man's pride and courage in order to understand why he didn't cry, as they say, on his friends' shoulders over the loss of Shakhmatovo, but instead just drily and firmly answered, "That's the way it goes." A brief note in his notebook tells us much more about his real feelings about the loss: "I had a dream about Shakhmatovo: ah. . . . " What does this "ah . . . " signify, what does it resemble most of all? I spent a long time trying to imagine this "ah . . . " in its natural form, said aloud, its degree of audibility and duration, and I concluded that most of all it resembles the cry of a person who has been unexpectedly injured.

Now let's turn to the very first source from which children of our time—that is, I mean, people of our time—learn that Blok had some sort of estate, and how he reacted to the loss of it. What can we do if in school we study Mayakovsky before Blok? We read:

> . . . During the first days of the Revolution I remember walking past the thin, bent-over figure of a soldier warming himself at a fire built in front of the Winter Palace. Someone called to me. It was Blok. We walked together to the Children's Entrance.
>
> I asked, "Do you like it?" "It's fine," said Blok, and then added, "They burned my library in the country."
>
> This very "It's fine" and "They burned my library" constituted two sensations of the Revolution.

The quotation is taken from V. Mayakovsky's obituary on the occasion of A. Blok's death. We read this obituary, of course, not during our school years, but much later when we encounter the collected works of Mayakovsky. But the thing is that, several years after the obituary, the revolutionary poet duplicated this scene in his poem *All Right!*[66] but this time in a rhymed form. And here it enters our consciousness, since who among us has not studied the poem *All*

Right! in school?

If we preserve the line breaks, the scene of the meeting between the two poets near the Winter Palace appears thus:

Holding
    his hands
        in the flames of the fire,
                a soldier
                    warms himself.
The fire
      struck
        the soldier's eyes,
lay
  on a tuft
      of hair.
I recognized,
    was amazed,
        and said:
"Hello,
    Alexander Blok.
The futurists are in luck,
             the tailcoat of the past
is coming apart
       at each seam."
Blok took a look—
          fires were burning—
"Very good."
Everywhere
      Blok's Russia
        was drowning. . . .

Strangers,
      northern mists
sank
  to the bottom,
     as fragments
          and pieces
of tin cans
   sink.
And right away
     a face
      altered,
        more wordless,
more gloomy,

    than death at a wedding:
"They write . . .

            from the country . . .

                    they burned . . .

                                    my
library at the estate."

In this manner, the incineration of the Shakhmatovo library (and it could, of course, only have burned down along with the rest of the house) became, after the example of Mayakovsky, a literary fact—and moreover a legend. Mayakovsky was a poet. The conception that had formed in his mind of Blok's seemingly dual relationship toward the Revolution was one that he clothed in a concrete literary form, just like his other conception as well, that Blok seemed to have perished along with Russia and seemed to find no place for himself in the new reality. We read in the same obituary from 1921: "I heard him in May of this year in Moscow: in a half-empty auditorium, silent as a graveyard, he quietly and sadly read his old poems about the singing of gypsies, about love, about the Beautiful Lady—there's nowhere for him to go in the future. In the future lies death. And it came."

Cruelly, though beautifully, said. Mayakovsky's words were born of the desire to affirm his speculative poetic conception, but nevertheless they stand in contradiction to pertinent facts. Of course, the fierce, militant (at that time) futurist wanted everything to happen to Blok in just this way; but what one wishes for doesn't always coincide with reality.

Eyewitnesses testify that Blok's trip to Moscow in May was a success: the auditoriums were packed to bursting, he was received with great enthusiasm, and the applause was deafening. Here, for example, is how Nadezhda Pavlóvich[67] remembers those evenings:

> The hall of the large auditorium of the Polytechnic Museum was overflowing. Young people crowded in the aisles. All the seats were taken. Many people held flowers in their hands. Venerable and not so venerable writers and actors stood crowded on the stage. Blok walked out onto the stage without ceremony. The hall shuddered in the emotion of the first encounter. . . . Afterward the applause was endless. People loved him very much and revered him as the first Russian poet of our time.[68]

The fact that Blok's poetry was not only alive and well, but loved too, is confirmed as well by the major poetry reading of Alexander Alexandrovich in St. Petersburg several days before the trip to Moscow. S. Alyansky's recollection of this evening is well known:

... Blok was scheduled to appear in a theater that held about two thousand people. ... A poster announcing the coming evening was put up all over town. ... At the theater's ticket office on the Fontanka a long line of young people had formed. They packed all the aisles in the orchestra section and the circles. ... With difficulty I made my way backstage. It was packed with people there, too, and the stairway was so crowded with people that the photographer who had come to take pictures ... barely got through with his cumbersome apparatus. ... It was a huge success. After each poem a whirlwind of applause and shouts rose in the auditorium. ... It seemed as if the public's enthusiasm would never abate. In the auditorium they had already begun to extinguish the lights, but the young people still weren't able to calm down.[69]

You must agree that this hardly resembles the depiction of a half-empty auditorium and graveyard-like silence.

In precisely the same kind of contradiction with reality is the vivid—I can't deny, even beautiful—scene of the meeting between the two poets by the fire near the Winter Palace in 1917, in the first days after the Revolution. If, by straining our interpretation to the utmost, we allow that Alexander Alexandrovich could still venture out onto the streets at night in a military uniform (I really don't know to what end), then he would have dressed specifically according to the standards of that uniform, according to all the strict demands of the uniform, down to the last neatly buttoned button. As is well known, he was never a soldier.

This is not the most important thing, however, but rather the fact that the Shakhmatovo house burned down, not in 1917, but in the summer of 1921, and the library did not burn at all, as we shall soon see; so, in 1917 Blok could not have talked with Mayakovsky about a library that had burned down.

After seeing Shakhmatovo soon after the fire in 1924, the writer Peter Alexeevich Zhurov wrote down the following in the words of the local peasants, in particular the words of Lukerya Yastrebova of the village Gudino (the manuscript is kept in Pushkin House):

The Beketovs and Bloks lived well and peacefully. The people of Gudino liked them very much. Sometimes they would go to Lizaveta Grigorievna or Lyubov Dmitrievna for medicine. The peasant men assembled to decide what to do with the estate. Residents of Gudino said to leave it the way it was, while those of Sheplyakovo wanted to sell it. Opinion was divided. The people of Gudino said, "We won't give it up!" But the others prevailed, and the president of the Executive Committee, Mazurin, supported their position. The annex and barn are now at Grigory's in Shemyakino, while the large granary is located at Volodka Usatovo's. Bolshakov carted away to his place at Shemyakino the little servants' house

that Blok had built. The bathhouse is in Gudino. The family icons are in Koto-
vo: they weren't sold at the auction [that means there was an auction?—V.S.].
The Vertlinsk District Executive Committee took all the books and papers on
three carts to the hamlet of Novoe. The table with the deep-blue tablecloth went
to the hamlet of Novoe. They smashed the piano and took it apart. They took
the white spruce and burned it. The entire house was demolished, but the walls
were joined with tenons and tarred—only they remained; the ceilings, floors—
everything was torn out. The rich made use of it; the poor were afraid to. They
burned down the house in the stubble-field in 1921: now there's rye in the field,
and there was rye then, too. The house burned like a candle.

The teacher at Tarakanovo knew very little about Shakhmatovo or about
Blok. Yes, they squandered the estate. There was a time when pupils at school
worked out problems on pieces of paper on the reverse side of which were
Blok's autographs. He brought me one of Blok's books: N. Minsky, *Religion of
the Future* (Religiya budushchego), published by Pirozhkov, St. Petersburg,
1905. The cover had been torn off. On the title page was the inscription:
"This book belongs to Vladimir Yastrebov, peasant of the village of Gudino. . . . "
In the book many places have been underlined and there are many notes in the
margins.

In Blok's diary of 1921 we find the entry: "In a little packet which Andrei
had saved from the house at Shakhmatovo, and which Ferol had brought to me
in fall, are pages from Lyuba's notebooks (a very large quantity). There was no
trace of her diary. Pages out of notebooks, pieces of my destroyed manuscripts, pieces of father's archive, notices, university notes (juridical and philo-
logical), assorted rough drafts of poems, and pictures that had hung on the
wall of the annex. On some of them there were traces of human hooves (with
horseshoes and dirt). That's all."

There is yet another place in Blok's article on Leonid Andreev:[70] "Today
nothing at all remains of these native places where I spent the best times of my
life; maybe only the old lindens still rustle, if they haven't torn the skin off of
them, too."

Regarding the library that was taken away, in the words of eyewitnesses, on
three carts, it is perishing in reality, but only without romanticism, not burning
in a purifying and fierce fire, but gradually dissolving in the cold and gray
gloom of indifference.

According to the words of Stanislav Lesnevsky, in the hamlet of Novoe the
former Shakhmatovo library was placed under the authority of the local coop-
erative, and books written in foreign languages were separated from the rest
and sent to the city of Klin. . . . P. Zhurov found part of the Shakhmatovo
library in the village of Merzlovo, where it had been stored temporarily in a

school. Subsequently these books were moved again to the hamlet of Novoe and stacked up on the veranda of a rest home. In 1926 part of the Shakhmatovo library was separated out for the rural library at Vertlinsk, while another part—the larger part—was transferred to the Klin Regional Distribution Center.

At this point all traces of the Blok library disappear.

Maria Andreevna Beketova conceptualizes the downfall of Shakhmatovo rather realistically: "In 1917 Alexandra Andreevna and I traveled to Shakhmatovo for the last time. After that it became impossible to go there, and soon the house was vandalized and burned down by the neighboring peasants—not out of malice, but just because, having set out to take care of the estate we had abandoned, they stole everything in the house over a period of time, and then wanted to hide the traces of their thievery."

Kopeikin, a local resident from Tarakanovo, related the following to Stanislav Lesnevsky: "They dragged away iron and threshing-machines. It was frightening there, at Shakhmatovo—at the edge of a forest, no one present, owls calling, eagle-owls. . . . We ran to Shakhmatovo—there were woven toys and pictures, and we were little, it was interesting for us, and so we asked for things. The estate was a wealthy one. . . . "

Another local resident supplements the picture: "When the ladies and gentlemen had left, the little kids and I dragged some pictures out of the manor house, there were many pictures in it. . . . One fellow was older than us and wouldn't let us into the house, saying: 'If you scratch the heels of my feet, I'll give you some pictures. And if you don't—you won't get any. . . .' We finished breaking up the piano, what a fine piano we turned upside down. At times we would jump on it with our feet. . . . We were little and didn't understand anything. . . . If only we had known. . . . "

The question arises: since Blok loved Shakhmatovo, why didn't he attempt to save it? He did live in St. Petersburg, his name was well known, and they could have reached an agreement with him, say, they could have issued him some kind of safe-conduct, and if he had just gone to Shakhmatovo himself, then it's possible that no one would have dared to wreck the house of a poet. After all, the peasant men told Maria Dmitrievna Mendeleeva after she had arrived at Shakhmatovo, "You know, if Lyubov Dmitrievna herself had come, she's the mistress of the house. . . . " But they didn't allow Maria Dmitrievna to enter.

There are many reasons for this, both psychological and entirely objective ones.

The main reason was that, in comparison with what had taken place in Russia, Shakhmatovo seemed a trifle, a little speck of dust. The cataclysm was so enormous and all-encompassing that, really, it would have been comical for Blok, who did not separate himself from Russia (if he had to perish, then it would be along with her; if he were to save himself, then it also would be with her), to think of saving some house with a mezzanine standing at the edge of a forest. And although from the purely factual perspective the meeting between the two poets at a fire near the Winter Palace was more than likely a legend, Mayakovsky with his poetic intuition was very close to the truth, only he altered the accent. Blok felt badly about Shakhmatovo, and he cried about it in his dreams; but he offered it as a conscious sacrifice to the Revolution, as he offered himself, and as he offered all of Russia. Of course, he had nothing to do with what happened to the Russia of the past, he could only accept or not accept it in his heart; but he was involved, at least to some degree and in practical terms, in what happened to Shakhmatovo. What can one say? This means that his sacrifice was offered all the more consciously.

> The poet was certain that Shakhmatovo and everything that was part of it had to perish—and in this loss he saw a necessary inevitability and retribution. . . . He couldn't set about saving his own belongings in the storm that he considered to be a purifying one. . . . Both A. A. and L. D. not only did not go to try to save Shakhmatovo, but in all likelihood even found it strange to think about something other than the events taking place at that time.[71]

Blok's own words testify to the fidelity of the above observation: "And that things are bad there [that is, at Shakhmatovo—V.S.], that things are bad everywhere, that the catastrophe is close, that horror is at the door—these things I have known for a very long time, I knew them even before the first Revolution. . . ."[72] "It is incumbent upon the artist to understand that the Russia of the past is gone and never again will return. The Europe of the past is gone and will not return. It may be that both will look ten times more horrible, so that it will become unbearable to live, but the kind of horror that took place will no longer return. The world has entered a new era."[73]

It is clear that in light of this kind of perception of the events it would have been comical for Blok to bustle about trying to save Shakhmatovo, or, as P. A. Zhurov correctly put it—[his] belongings.

Moreover, one has to imagine the general circumstances of those years in order to understand that even if they had wanted to, in fact neither Blok himself, nor his wife, nor his mother was in any condition to think about Shakhmatovo during those years.

The cold and the famine. Ration cards. The struggle for the most elementary kind of existence. When a person is cold and hungry, his or her thoughts and desires are constrained, activities are halted, and interests deadened.

And life changed radically. Let's take a look at only two lists, two little registers, two columns of numbers seemingly dispassionate and dry, but we'll see that underlying each little list is a way of life, life itself, in all its forms of well-being or in all its horror. You see, the Bloks always kept records of their expenses. Their little notebooks in small format with a fine binding survive today in Pushkin House. We may be amazed, of course, at the coexistence in one person of a gigantic poetic soul and a predilection for the scrupulous recording of expenses, but here it's evidently a matter of his upbringing, of habit, and of family tradition, if you will, rather than of the deep-seated characteristics of his soul. And so, let's leaf through them quickly and quote at random several entries in the expense books for Shakhmatovo. To save space on the page, I'll write these notes, not in a column, as they were recorded in the notebook, but from left to right, even though they would appear much more stark if presented in column form.

> The cow Lysenka: 95 rubles. Coupling-bolt: 50 kopecks. Village constable: 2.62 k. Seedlings: 1 r. Ploughshare: 20 k. Cart: 1 r. 50 k. Horse: 124 r. Small shot: 1 r. 70 k. Two barrels: 65 k. Ten pounds of clover: 3 r. 63 k. [Of course, this means clover seeds—V.S.] Flower beds: 60 k. Five days' harrowing: 3 r. 50 k. Three day laborers: 85 k. Stove-setter: 1 r. Two peasant women for two days each: 1 r. 60 k. Tips for the housepainters: 40 k. Transportation of the sheaves: 40 k. Drying of oats: 70 k. Chopping of ice: 3 r. Day laborer brought ice for two days: 40 k. Postcard: 3 k. Ananievna for January: 4 r. Nikolai: 25 r. Telegram: 59 k. Wheel grease: 1 r. 50 k. Rope: 1 r. 80 k. Tar: 1 r. 60 k. Osinki peasant women for harvesting: 1 r. Four rakes: 2 r. 40 k. Two scythes: 3 r. 20 k. Pitchfork: 75 k. Horseshoe: 15 k. Peasant women for drying: 3 r. Tea and sugar: 2 r. 50 k. . . .

And so on and so on. One could just as well copy out all of the expense books of the Bloks.

In the same Pushkin House are housed some individual notes of Lyubov Dmitrievna, and no longer in a little book with a fine cover, but on gray, coarse-fibered packing—in a word, paper. And, moreover, on pitiful scraps of paper. I've copied out several of the numerous notes, once again violating the columns: Rags and clothes: 10,000,000 r.; Brussels shawls: 1,500,000 r.; Scarves: 15,000,000 r.[74]

This, of course, does not refer to the acquisition of shawls and clothes, but to the selling of them. Concerning the purchases, the lists look like the following: "Boiled potatoes: 3,000,000 r. Wood: 10,000,000 r. Medicine: 1,035,000 r. Doctor: 5,000,000 r. For the apartment: 1,500,000 r. Milk and cranberries: 1,320,000 r. Eggs: 2,500,000 r."

There are also lists of things they intended to sell. I copied out the following, skipping from number to number:

18. Mom's broken lamp. 19. Auntie's empty dresser. 20. Theater collection. 21. Rags and clothes from the trunk. 22. "Le théâtre." 23. The green dining-table chair. 24. Room divider of red wood. 27. Auntie's walnut bookcase. 28. Mom's wardrobe with a mirror. 29. Provisions hamper. The carton with the military badges. Walking sticks. Military boots. Collars and cuffs. Secretary desk with unstable legs. The small sofa with the latticed back. . . .

And once again we find: "Potatoes: 7,000,000 r. Wood: 10,000,000 r. Medicine: 5,000,000 r."

Let us admit that, in light of the way of life that emerges from these registers, these people would not have been concerned about the little house with a mezzanine that remained far away at the edge of a forest but in fact in another world, even virtually on another planet, open to all the winds and all kinds of misfortune.

Blok never again returned to Shakhmatovo, but he did see it, and many times. Well, first he saw it more than once in his dreams. In his notebooks we read: "I dreamed of Shakhmatovo: ah . . . ." "Why did I cry so hard about Shakhmatovo in my dreams last night?" "Dreams, dreams once again: especially about Shakhmatovo. . . . "

Second, almost until his death, as late as July of 1921 (he died in August), Blok was working, in a weakening hand, on new lines for the poem *Retribution*, seizing with an imagination now faltering the scenes dear to his heart. It cannot be ruled out that Blok himself was growing weaker while his imagination was functioning in an intensified manner, showing everything vividly and distinctly. And these scenes that he immediately transferred to paper were scenes of Shakhmatovo. So without exaggeration we can say that Blok lived and breathed Shakhmatovo until his last breath, until the last beat of his heart, until the last pale-blue flame in his fading brain.

✦ ✦ ✦

Enormous silvery poplar
Bent its canopy over the house,
The yard received each visitor
With a wall of fragrant dog rose.
There was a granary with peaked roof
Protected from northern winds,
And one could clearly hear

The silence as it bloomed and slept.

✦ ✦ ✦

And the gray house, and on the mezzanine
A Venetian window
Of colored glass—red and yellow and blue,
As if that's how things ought to be.

✦ ✦ ✦

The sun casts shadows of the leaves,
And the wind outside the window bends
The hundred-year-old lilac bushes,
In which the old house is drowned.
There's a muffled sound—
The sound of that same silence,
Or church bells in the distance,
Or the rumble of (unfinished) spring.

✦ ✦ ✦

And the ringing door of the balcony
Opened out onto lindens and lilacs,
And the deep blue cupola of the sky,
And the indolence of distant villages.

✦ ✦ ✦

The church shines white above the river,
Beyond it once more lie forests and fields . . .
And in all the beauty of its spring
The Russian earth is glowing.[75]

These lines were written only from memory. At this time there was noth-ing left at Shakhmatovo. Even Shakhmatovo itself, as such, no longer existed. Blok outlived the estate by about two weeks.

The forest that surrounded the Shakhmatovo property on all sides and had been controlled by human activity—a cultivated garden, axe, saw, spade, hourly surveillance, flower gardens and flower beds, clearings and paths—this forest stirred from all sides like a united army or, more precisely, like a crowd, and swallowed the Shakhmatovo clearing as if green waves had closed over it.

IV

At least forty years passed in this way. Various events took place in the country, its way of life and character changed, and the consciousness of the people changed, too. Collectivization, the construction of canals, five-year plans. The little villages began to disintegrate, the various evening bells one might hear fell silent, and the little churches themselves (gleaming white on neighboring hills) were snuffed out like little flames in a damp wind. Songs began to ring out about Katyusha, and Kakhovka, and a vehicle on springs. Tractors in the fields and airplanes in the sky began to rumble. Twice a war rolled back and forth over the land, while the Shakhmatovo clearing slept and slept, engulfed by the green wave of the forest. And, I reiterate, about forty years passed in this way.

Well, it almost seems as if Peter Alexeevich Zhurov's visit doesn't count. It took place "hot on the trail," in 1924. It was still possible to discern traces of the garden, even though it had begun to look wild and overgrown. According to Zhurov, still visible were the charred soil, pieces of brick, fragments of the window casings, daisies on the lawn, a semicircle of lilacs and acacias, traces of the paths ("their motif was one of winding and curving"), and a sofa created by the sod. . . . But this was still charred ruins, a wound not healed by nature, and subsequently decades passed from the day of the fire: until 1931—the first decade, 1941—the second, 1951—the third, 1961. . . .

After the war Viktor Sergeevich Molchanov worked as a photographer for the Literary Museum in Moscow. His immediate responsibilities most likely consisted of making various photocopies of documents, manuscripts, and old photographs, and also of recording the literary events that took place in Moscow—meetings, evenings, and funerals—so that a photo library would be built up in the museum. However, being a hard worker and an enthusiast, Viktor Sergeevich did not limit himself to these responsibilities. Preferring to all other means of transportation a bicycle with a little motor adapted to it and piled high with camera equipment, he would make his way to all the obscure places in the environs of Moscow and its contiguous provinces, along washed-out roads, ruts overgrown with grass, and forgotten paths, in search of well-known, but sometimes forgotten, literary places. This man's character can be explained best of all by his answer to a question on a semiserious "Alepino questionnyaire" [*sic*—V.N.]—namely, "What is your favorite kind of physical work?" Usually people answer, "Mowing hay, splitting wood," etc. Viktor Sergeevich artlessly wrote: "Hauling heavy loads."

Tyutchev's Muranovo, Chekhov's Melikhovo, Prishvin's Dunino, Paustovsky's Meshchera and Tarusa, Esenin's Ryazan, and Zakharovo and Serednikovo near Moscow, which are linked with the names of Pushkin and Lermontov, subsequently became the subjects of Viktor Sergeevich Molchanov's photo-artistic, conscientious, and painstaking investigation. He did not strive for the level of artistic expressiveness that perhaps characterizes contemporary photography, in which a single branch in a close-up against a washed-out background creates a virtual picture. Or when a group of birch trees is photographed from below looking up, and what develops is like the cupola of a cathedral. The position of a museum employee imposed additional responsibilities: aside from all its other qualities, the photograph had to be a document. But they turned out to be beautiful and expressive anyway, due to the force of the possibilities, the taste of the master, and his love for Russian literature and nature in the Russian countryside.

Hence, this same Viktor Sergeevich Molchanov, who loved to haul heavy loads, indeed made his way to Shakhmatovo on his bicycle before anyone else, after four decades of total neglect.

Viktor Sergeevich returned from this trip as if he had dived into the depths of the ocean with an aqualung and had found signs of the submerged Atlantis there. Familiar names that had been preserved, it seems, only in scholarly works on Blok acquired a reality: Gudino, Osinki, Praslovo, Runovo, Boblovo, Tarakanovo. . . .

The first thing we did was rush to look at the photograph that recorded the ruins of the church at Tarakanovo. After all, if there was a tree in the picture, then we still had to guess whether or not Blok had known this very tree, or whether it had known Blok; if there was a tree stump, then was it left from that same poplar that had overhung the house?—we could only guess about all of these things. The church was the original one at Tarakanovo, the same one in which the bride and groom had been married, which Dmitry Ivanovich Mendeleev had entered, near which troikas with little bells on them had crowded, out of which the coffin containing the body of Professor Beketov had been carried, in which "a young woman was singing in the church choir. . . . "

In an old photograph that had survived, which was taken by an unknown person presumably in the 1920s and was sent to me by Stanislav Lesnevsky, the church was still completely whole and intact. The little cross was there, and the cupola, and the small belfry next to it. There was no background: only a clear, white sky. The photograph is authentic but not expressive. That is, it is expressive in the expressiveness of the architecture itself, but no more than this.

Molchanov brought back a photograph full of atmosphere, trepidation, and horror. The sky was filled with cumulus and black clouds, with a bright ray of light right behind the church. The ruins are prominent and serve as the subject. One sees a pile of road-metal near them, stinging-nettles, roots of a dead tree, gaping window-openings, exposed holes in the brickwork. . . .

We were also agitated by a photograph of three birches, old ones, obviously Blok's, all that remained of a long lane—more precisely, of the line of birches that at one time had marked the entrance to Shakhmatovo from the side of Podsolnechnaya Station.

Molchanov said that at Shakhmatovo it was difficult to determine today where everything had been: the garden, the pond, the house, the annex—everything had been eroded and become overgrown. The very views that formerly could be seen from the Shakhmatovo hill were closed off by the forest, as if by a window blind.

Molchanov showed us these photographs in the studio of the artist Glazunov. Immediately we began to read Blok's poetry, entered into conversation, became excited, and agreed to make a trip to Shakhmatovo. At present I don't remember, after almost fifteen years have elapsed, what prevented me from going, but Glazunov and Desyatnikov (the art historian) made the trip and brought back a small table of Blok's that they had found and pried away from an old woman. This old woman seemed to remember Blok well, had carried mail to Shakhmatovo as an adolescent, and had worked there as a day laborer, washing floors and clothing, and digging flower beds.

Soon afterward Ilya Glazunov published in *Literaturnaya gazeta* (in 1965) the article "Tam, gde zhil Blok" (Where Blok lived) and two of his drawings. These were the first words about Shakhmatovo published in a newspaper.

The stories of Glazunov and Desyatnikov excited me even more. I wanted to go there, meet with this woman and question her, at the very least take a look at her (she took mail to Blok!), but my daily routines carried me along, whirled me around as if in a whirlpool, taking me now to the shore, now toward the rapids, now driving me farther from my goal, now bringing me closer to it; and meanwhile a new actor appeared in the story of Shakhmatovo. I wrote a brief afterword to one of this man's books. From it I cite several sentences that immediately will provide the necessary introduction to him:

Stanislav Lesnevsky is well known in literary circles and to those who love literature as a literary critic who writes mainly about poetry, who keenly understands it, and who follows its development.

The work of most critics fits into the usual boundaries and genres of this profession: reviews, surveys, and literary articles; but the devotion to their native land, glorification of it or reflections about it, about its fate and directions, seems somehow the domain of poets and prose writers. However, a feeling for one's native country is characteristic of every person, after all, regardless of his or her profession. In the present case, the critic and literary historian Stanislav Lesnevsky is richly endowed with this feeling. As someone who writes, he didn't want to leave his observations, feelings, and thoughts outside the bounds of his creative attention. These articles and essays appeared as the result of his behavior and activities in the study (and at times, in the elucidation) of literary memorial places. His behavior is called "active" because he has not just traveled to, looked at (studied), and written about, but immediately included himself in the battle for the preservation and restoration of the memorial appearance of this or that notable place.

Following in the footsteps of Molchanov, Glazunov, and Desyatnikov, Lesnevsky also made his way to Shakhmatovo, Tarakanovo, Boblovo, and Solnechnogorsk (the hamlet formerly called "Solnechnaya Gora"), but, in contrast to the previous visitors, he energetically began to act. Starting from the end of 1968 he collected signatures to a letter, and on July 23, 1969, published this letter in *Literaturnaya gazeta*. Here is the letter, along with the signatures to it. It bears almost the same title as Glazunov's article, "Zdes zhil Alexander Blok" (Here lived Alexander Blok):

There are sacred places that in a people's memory have converged with names constituting the pride of its country's literature. Mikhailovskoe, Yasnaya Polyana, Spasskoe-Lutovinovo, Karabikha, Melikhovo, Muranovo, Konstantinovo. . . . These and other notable corners of our Native Land evoke images in our imagination which are dear to each of us, which have not faded with time. Literary-memorial homes-museums, country estates-museums—these are, you know, not only monuments of the past, but also living hearths of today's culture. Everyone knows well enough how much we are doing for their preservation.

It is all the more evident how large our debt is to those glorious places that are inseparable from the name of the great Russian poet, one of the outstanding founders of Soviet literature, Alexander Alexandrovich Blok. We believe that it makes sense to consider the question of the possible creation of Blok museums in Leningrad and in the environs of Moscow—in places linked with the life and works of the poet. . . . Into Blok's poetic consciousness there entered impressions imbued with the estate located near Moscow, Shakhmatovo, and the nearby villages of Gudino, Tarakanovo, and Boblovo, with their beautiful natural setting. It is precisely here, at Shakhmatovo, on the spot of the former estate or near it, that a Blok museum located near Moscow could develop. It would be desirable to declare the places Blok knew as sacred, protected, and memorial. . . .

The immortalization of the memory of a great poet is our patriotic duty.

The letter was signed by the following: M. Anikushin, R. Gamzatov, S.

Konenkov, A. Prokofiev, S. Richter, G. Sviridov, A. Tvardovsky, G. Tovstono-gov, G. Ulanova, K. Chukovsky, and D. Shostakovich.

What a constellation of names! Every single one is a laureate of the Lenin Prize. And so, did they themselves, while at the same time living in different cities, all get together and say, "Let's compose a letter!"? Well, no. It's done like this: the initiator composes the letter and collects signatures by calling on the desired cultural activists; he or she goes to their apartments and their dachas. They read the letter and, if they agree with its text, sign it.

After the publication of this collective letter, Stanislav Lesnevsky published a rapid-fire series of articles in various newspapers: "Chtob byli pamyatyu khronimy" (So they will be preserved in our memory), *Moskovskii komsomolets*, October 11, 1969; "Pamyat trebuet" (Memory demands it), *Sovetskaya kultura*, October 30, 1969; "Veet strokoi Bloka" (The wind blows with lines from Blok), *Literaturnaya gazeta*, January 14, 1970; "Zdes dolzhen byt zapovednik" (There should be a preserve here), *Literaturnaya Rossiya*, March 6, 1970. (The last article is signed by P. Antokolsky and V. Orlov, but S. Lesnevsky organized it and provided the commentary to it.)

After this, a chain reaction occurred: A. Levina, "Zdravstvuite, Alexander Blok!" (Hello, Alexander Blok!) (an excursion through the future preserve), *Komsomolskaya pravda*, July 19, 1970; G. Borina, "I vechny boi . . . " (And the eternal struggle . . .) (notes from the exhibition dedicated to the ninety-year anniversary of the birth of A. A. Blok), *Znamya Oktyabrya* (Solnechnogorsk), August 29, 1970; Yu. Dorokhov, "Den poezii v Shakhmatove" (A day of poetry at Shakhmatovo), *Leninskoe znamya*, August 11, 1970.

Here are the titles of some other articles that followed, by various authors: "At Shakhmatovo," "Invitation to Shakhmatovo," "His Russia Is Our Russia," "The Voice of Readings at Shakhmatovo," "On the Road, Open to All Eyes," "The Russian Land Glows in Its Beauty," "I Didn't Seek a Better Fate," "May the Nightingale Garden Bloom," "The Places Blok Knew," "Seventeen Versts from the Station," "And I See a Pale-Blue View," "And a Link with the World Was Affirmed," "The Unfading Flame of Poetry," "To Care for the Inimitable," "Memory Needs a Refuge," "They Lived at Shakhmatovo," "The Resonant Voice of a Poet," "Blok's Garden," "A Meeting with Shakhmatovo," "Alexander Blok. Links of the Memory. . . . "

In a word—an avalanche of articles, notices, reflections, entreaties, appeals to finer feelings, reportage from the site of the events.

But you won't move anything from its place by articles alone. You have to go there yourself, win people over, write papers, make demands, and act.

Stanislav Lesnevsky joined the Jubilee Blok Commission (he initiated its creation) of the Moscow Writers' Union, and then the city committee, regional committee, regional department of people's education, provincial department of people's education, the Ministry of Culture—whatever you like—and no longer was received as a private citizen but as a representative of the Jubilee Commission: at that time, the ninety-year anniversary of the poet's birth was approaching. In 1971, in front of School No. 1 in Solnechnogorsk, a bust of Alexander Blok was installed and ceremonially unveiled. At this school, which already had borne the name of Alexander Blok prior to this, there was a Blok corner. It contained a few photographs and several books. The thought occurred to Lesnevsky to set up a large photo exhibit about Blok, and not in Solnechnogorsk, but closer to Shakhmatovo. Of course, one can't set it up in a forest. Nearby villages, such as Osinki, were also not suitable. One can't very well set up an exhibit in a country *izba* that is already inhabited. Tarakanovo proved the best suited for it. And, conveniently, a village elementary school was being vacated there. The school had been established long before by the zemstvo, around the year 1900, and had taught more than one generation of little kids to read. Alexander Blok, too, knew this school. By 1970 the predicament of the village had become such that there were no longer any children left to teach. This was a typical scenario for provinces in the region of Middle Russia: Tula, Vladimir, Ryazan, Smolensk, Yaroslav, Ivanovo, and many others. In each province several dozen schools would close every year.

The school in Tarakanovo, too, had closed. An event that was not inherently cheery couldn't have turned out to be more serendipitous for the idea of setting up a photo exhibit. The exhibit was mounted. An article about it appeared in the Solnechnogorsk newspaper:

Everyone who has recently visited the newly famous Shakhmatovo cannot have missed the photo exhibit devoted to A. Blok. . . . The exhibit is situated in an amazingly appropriate place, in a modest building on the edge of the village of Tarakanovo, where formerly a country school had been. Alexander Blok used to come here. And the exhibit has turned out to be organically linked with the great poet's name, with his life and works. . . . The photo exhibit includes approximately three hundred documents, copies made from the originals at Pushkin House and the Museum of D. I. Mendeleev in Leningrad, the Literary Museum. . . . The exhibit, which was assembled by the Jubilee Commission of the Moscow Writers' Union, helps us to genuinely understand and appreciate the works of A. A. Blok. "The contact with the beautiful purifies the soul, and these amazing expanses all around prompt one's hidden feeling of love for one's country to surface from the deepest recesses. We thank the organizers of this exhibit." This is one of the numerous entries in the journal of comments. . . . The exhibit is very interesting, but it's a great pity that it will be open for only a few more days. The writer Lesnevsky is leaving for Moscow, but a replacement for him hasn't been found yet. And it's not clear what will happen to the exhibit. The district cultural office has not yet decided its fate. And they really could transform the exhibit into one that is permanently open, after establishing a village library in the school building and training public tour guides from among the cultural workers. We cannot deprive Blok's homeland of this splendid exhibit.

Everything turned out exactly as it was described. A village library was established at the former school, and the exhibit became permanent. It remains open to this day.

However, distant shores such as Solnechnogorsk and Tarakanovo, in which one could consolidate one's grip, could not satisfy the fanatical zeal of the patriot. Tarakanovo is still not Shakhmatovo, but only the approach to it. At Shakhmatovo itself there was only a forest and a clearing—there was neither a bust near the school, as in Solnechnogorsk, nor a photo exhibit, as at Tarakanovo. No foundation, or even traces of one, remained of the house. What could be done? Stanislav Lesnevsky and the artist Yury Vasiliev walked through the clearing at Shakhmatovo together. What could be done? What could be done? There was nothing there that would catch the eye. And yet in August (1970), on the first Sunday of August, a large literary celebration had been scheduled to take place here. It had been announced, and people had been invited from Moscow, Solnechnogorsk, hamlets, and villages. Let the people gather, let the writers and poets come, let their words about Blok and poems by Blok ring out, that's all well and good. But where? How? At which spot should they read? Where could the public stand? On the spot of the former house there was impossible brushwood: trees, bushes, and stinging-nettles. The clearing was some distance away, but it was without character, unorganized. Just imagine: people would gather, and they would search for at least some object to catch the eye, their consciousness, their hearts. It couldn't very well be a withered tree stump among the nettles or a little sign on a pole, "On this spot stood the poplar that overhung Blok's house," could it?

Yury Vasiliev walked and walked, lost in thought, and suddenly his face lit up.

"I've got it. We need a stone."

"What kind of stone?" said Stanislav to the artist, not understanding at first.

"A large one, a fine one. A point to focus on. It should be brought from somewhere and set down in the clearing. A stone is something solid, inspiring, and stable. We have to search for one."

Their first thought was to find the stone near Runovo, on which Blok started writing his poem *Retribution*. That would have been just marvelous. They began to question the local residents.

"Near Runovo?" reflected an inhabitant of Osinki, upon being asked. "I don't know of one near Runovo, but there is a 'Holy Stone.' Only it's not near Runovo but by the forest at Praslovo, in a ravine."

"Why is it holy, what do you mean, 'holy?'"

"Maybe it fell from the sky. Or maybe people prayed near it in olden times, before churches existed. They say that it grows. . . . "

"What grows?"

"The stone. From century to century it keeps growing bigger. You can't verify this, of course, because we only live in one century. . . . But the stone is large, even enormous."

"How do we find it?"

"You see, you cross the oat field and keep to the edge of the forest, then you'll reach a ravine. . . . No, I don't think you'd find it. Come on, I'll show you. . . . "

The stone turned out to be the king of all stones. It lay on the slope of a ravine that was covered by forests. Such woody ravines are still called "gullies," at least where I come from, Vladimir Province. It's difficult to imagine that, in such a dense forest with bushes, people in pagan times would gather around this stone, as the peasant from Osinki had conjectured. But it's also true that on this spot in the distant past there could have been a bare slope, level, green, and inclined toward the sun. The face of the land changes very quickly. Lev Tolstoi was buried in a clear, high place with a marvelous view in all directions, but now his grave is in a forest. And a century hasn't even passed.

They walked around the stone, looked it over, and noticed that on one side its surface was furrowed with deep creases that, by exerting a certain amount of imagination, one might think formed the features of a human face, as if the stone had been lying here on the ravine's slope, century after century, having tormented thoughts about something.

"About ten or twelve tons," estimated Yury Vasiliev with his artist's eye.

Stanislav was thinking about the practical side.

"I can't envision which organization we'll have to contact in order to move this stone to Shakhmatovo. And how much will it cost? And for how long will they dawdle? And who will give us the money for it?"

"It's completely hopeless. Not a single one of the organizations that I can recall at the moment would undertake such a task. Money—without a written order. That means we need two organizations. One to do the work and the other—to pay for it. The search for these organizations, correspondence between them, various agreements and official registrations, determination of an estimate, and the execution of the work itself will drag on for weeks, if not months. And in fact, which organization, if not the Writers' Union, would take upon itself the transportation of this stone? And we still have to convince them of the expediency of moving it. The practical aspect of it. After all, what we've dreamed up is pure lyricism. It would be easier to pave the way for the restoration of the house at Shakhmatovo than for the transportation of this stone. No, it's completely hopeless."

"Wait a minute. Do you remember we saw a truck crane on the Tarakanovo Highway? They're doing some kind of work there."

"What of it? You think they're going to drop what they're doing right away and drive ten kilometers in order to haul this stone . . . ?"

"Maybe they'll do it. They're living people. There might be three or ten of them. . . . We wouldn't be losing anything."

By evening of that same day, the stone (it actually did weigh twelve tons), swinging on the overburdened boom of the truck crane, stirred from its many-centuries-old place and floated into the air. Then it made a short trip through a field road and was carefully lowered onto the grass, onto little summer flowers on a sloping clearing that resembled an amphitheater. Whether or not pagans had prayed near it, from this day on it was destined to serve as the platform for poets and anyone else who wanted to say something about Blok, read poems about Blok, or read Blok's own work. The "Holy Stone" became Blok's stone, Shakhmatovo's stone, the stone of memory and poetry.

It is unknown for how many centuries (millennia?) this stone had lain on the slope of the woody ravine, in its primordial place; and it is unknown how long it is destined to lie here, and whether at some new, distant turn of history it won't be the case that, in place of a heathen temple, in place of the memory of Blok, it will signify something new, take on a third meaning that we at present cannot imagine, as the naive pagans could not have imagined its current purpose.

We succeeded in stirring up the elements. On August 9, 1970, buses and automobiles made their way along the Tarakanovo Highway. From the village of Tarakanovo it was no longer cars (they remained in Tarakanovo) that made their way along roads and paths through the forest (about three kilometers), but rather files of people dressed in bright and festive clothing. The residents of Solnechnogorsk, Klin, Zelyonograd, surrounding villages, Muscovites, and ultimately several thousand people gathered for the first time at the place where Alexander Blok's house had stood. Among the local residents there were many little kids, children of various ages; and, as regards the townspeople, women were in the majority, as is the case at evenings of poetry, in theaters (and, incidentally, in church), concert halls, and the conservatory. For some reason, at all such events women seem to be more active (more spiritual?) than men.

And thus, thousands of people were gathered. What would they have circled and seated themselves around in this place if in this clearing there had not lain a large stone, in its appearance resembling a monument, or at any rate an installed monument?

After walking through the village of Osinki, through the wide, gradually sloping oat field and the dark forest, the people found themselves still in a forest with poorly marked gaps in it. To be sure, the presence of lilacs and dog-rose bushes in and of itself indicated that there had been something here at one time; however, after wandering for a while through brushwood, the people walked on, as if looking for something more substantial than a lilac bush or stinging-nettles, and at last they came out onto a more spacious and clean clearing with the stone in it—something to catch the eye—and they began to swirl and gather in groups around it, like a homogeneous, numerous swarm of bees swirls and gathers in groups around its queen.

If memory doesn't deceive me, during the first festival, poets and writers gave their readings right from this stone, but subsequently they always set up a temporary wooden platform near it, akin to a stage.

And hence every year people travel to this clearing, and Blok's poems resound, and all of this really does constitute a festival of poetry.

Stanislav was successful in paving the way for yet another matter. He agitated at the Moscow Bureau of Excursions (to young tour guides and enthusiasts), and soon they began taking tourists to Shakhmatovo to see the places associated with Blok.

"But, you know, there's nothing there," said the skeptics. "What can one show there? Nothing but a forest, and some stone. . . ."

The fears of the skeptics were not entirely unfounded. If they had wanted to, they could have gloated later on, when one day a dissatisfied tourist wrote a letter to the editor of the magazine *Turist*. But it's true that this letter was the only one during this entire period, while many encouraging and positive notes were written in the book of comments. . . . But perhaps some other people were dissatisfied as well: after all, not everyone who is dissatisfied will immediately and without fail write a letter!

The tourist's letter and necessary response to it further illuminate the subject at hand in two different ways, and moreover I can offer this correspondence here as a curiosity. The letter is entitled "How We Made a Tourists' Excursion to Shakhmatovo" (Kak my ezdili v turisticheskuyu poezdku v Shakhmatovo). I'll cite it below, preserving its style and other particulars, down to the transformation of the hamlet of Tarakanovo into the village of Tarakanovka, and likewise not quibbling along the way with the categorical and absurd pronouncement to the effect that "the marriage in it [the church—V.S.] of A. Blok to the daughter of D. I. Mendeleev is no reason to. . . " etc. Well, you'll read all of this for yourselves now:

On Sunday at 8:00 A.M. the bus left its starting point with tourists and sightseers; it was filled with those desiring to become acquainted with Shakhmatovo, the summer home of the author of one of the revolutionary poems, *The Twelve*— Alexander Blok.

The bus turned out to be unsuitable for a trip out of town and for tourists. It was one of those unfortunate buses intended for short drives on the bustling streets of the capital. There weren't any seats on the platform near the rear door. The entire platform was for standing passengers. Some of the seats faced the rear of the bus. For this reason it was difficult to hear the tour guide, and we had to stand up in order to hear his explanations. This was inconvenient. There is a rule stating that buses going out of town should not carry standing passengers, but only seated ones. With the abolition of conductors this rule is no longer being followed. It was passed with the aim of protecting the lives and health of passengers while en route, in the event of sudden stops of the vehicle, sharp turns, etc.

Fortunately, several people who had signed up did not show up, and we were able to arrange ourselves in the seated areas.

The road to the village of Tarakanovka, along the Rogachyov Highway, was bad, and the bus shook severely; it was good that no one had to stand. There were enough seats for everyone. But everyone had to hold on to keep from falling. The driver explained that his vehicle was not intended for out-of-town trips. In traveling by bus in Moscow Province, and even to Novgorod and Oryol, we haven't had to make excursions on such unsuitable buses.

After arriving in the village of Tarakanovka, we visited the photo exhibit in three rooms of the local library. Everything that was shown and explained there can be found in books about Blok. One doesn't necessarily have to drive over sixty kilometers, along a highway in poor condition, in order to see this exhibit. Not far away stood the ruins of the local church. It was demolished back in the 1930s, when "zealous" antireligious types were destroying historical monuments of church architecture, starting with the Cathedral of Christ Our Savior in Moscow. Ultimately the church was brought to a state of ruin during the war years. The front passed not far from there in 1941. There's no reason to restore it. In and of itself, the event of the marriage in it of A. Blok to the daughter of D. I. Mendeleev, L.D., still doesn't constitute a reason for its restoration. But to clean up this place during an unpaid Saturday[76] under the supervision of a Moscow historian would not be difficult. In one day a force of about twenty people, with the help of machinery, could clear away the bricks lying around everywhere, fill in the holes in the ground, clean up the area around the little square, and restore order to the existing graves and determine who was buried there. And besides this, of course, they should construct a temporary roof, if only out of tar paper, and hang up a board with the usual inscription, "Protected by the government—an architectural monument of the eighteenth century." And that's enough! There's a pond nearby, and they should install a couple of benches and plant a lilac bush in memory of the poet, and a bird-cherry tree, etc. In the village there are many such bushes, in every front garden. The townspeople won't pick all the blossoms on lilac bushes. The way the church looks right now is a scandal! It's a disgrace to the Moscow Provincial Department of Culture; it indicates that the law of the preservation of the cultural and historical monuments of Moscow Province is an ordinary piece of waste paper, and nothing more. It indicates that these laws are not for Moscow Province.

Later on the question arose of how to get to Shakhmatovo. The tour guide explained that, besides the names of places, there's nothing there! The house burned down back in 1921. Everything that remained was subsequently dismantled and dragged away. Except for a single poplar, there's absolutely nothing there. There's no road, either. A path leads from the village through a ravine and forest.

So where's Shakhmatovo? It doesn't exist! Why organize an excursion? Why collect money from tourists when there's nothing to see? For what purpose? Is this how an excursion ought to be organized? Is this how Soviet poets ought to be memorialized?

Why is it, then, that in Dyutkovo, the village four kilometers from the Spaso-Storozhevsk Monastery, twelve kilometers from Zvenigorod, it was possible to create the wonderful museum in memory of Taneev and Levitan? Why?

I ask the Provincial Department of Excursions and Tourism to turn its most serious attention to such an excursion. The excursion to Shakhmatovo should be eliminated. The Division of the Provincial Department of Culture is advised to clean up the church at Tarakanovka immediately.

There is a signature, but I shall refrain from including it, since the letter wasn't addressed to me and I would need the author's permission in order to publish it. The official and well-founded answer of the Bureau of Excursions looks like this:

The Board of Directors and Methods Division of the Moscow City Bureau of Excursions, having examined a copy of the letter sent to the Central Soviet by the editor of the magazine *Turist* concerning the advisability of a tourist excursion to Shakhmatovo, report the following:

The excursion to Shakhmatovo was established in 1970, the year when a widespread celebration took place in our country to mark the ninety-year anniversary of Blok's birth. The excursion was developed very meticulously, with the participation of employees from the State Literary Museum and members of the Moscow Writers' Union. The Bureau of Excursions was most receptive to the numerous requests of Muscovites to learn about the memorial places in and around Moscow that are linked with the name of the poet who glorified the Revolution and Russia. Two excursions were created: "A. Blok in Moscow" and "A. Blok in Shakhmatovo."

The excursion to Shakhmatovo takes twelve hours. It includes the showing of the vast region northwest of the area of Moscow linked not only with the name of A. Blok, but also with that of his grandfather—the important Russian botanist A. N. Beketov—and of the brilliant Russian scientist D. I. Mendeleev. We show the entire territory, which is permeated with the genius of these great people. It includes the villages of Osipovo, Vertlino, Sergeevka, Tolstyakovo; it also includes the Tarakanovo and Rogachyov highways, the Praslovo Forest, Tarakanovo, and Shakhmatovo.

In the hamlet of Tarakanovo, in the building of the former zemstvo school (a building from Blok's time, under governmental protection by a decree of the provincial Soviet of May 31, 1977), the only exhibit of its kind, devoted to the life and works of A. Blok, was mounted. Such an exhibit is not available in either Moscow or Leningrad. The exhibit was organized by the State Literary Museum on the highest level of museum work, with the intent of publicizing these places. In order to see this exhibit people come not only from a distance of sixty-five kilometers outside of Moscow, but from the most diverse corners of our country, to which entries in the visitors' registration book at the exhibit and in the book of comments eloquently attest. The following well-known writers have written more than once about the exhibit: Hero of Socialist Labor K. M. Simonov, Laureate of the State Prize P. G. Antokolsky, A. Turkov, Z. Mints, etc. All of them made entries in the book of comments about the exhibit.

The only monument still in unsatisfactory condition is the church in the hamlet of Tarakanovo, in which A. Blok was married to L. D. Mendeleeva (it was taken under governmental protection by the same decree of the provincial Soviet of May 31, 1977). In 1976 restoration of the church was begun (the restorers are L. A. David and V. I. Yakubenya of Moscow Provincial Restoration Studios). A museum in the church will be open for the celebration of the centennial of the poet's birth.

This anniversary will be celebrated in 1980. A special commission has been created, which has outlined an entire series of major undertakings to perpetuate the memory of the poet. The project to restore the country estate at Shakhmatovo has been confirmed. Not to show these territories on the basis of a church's being restored there seems ill-advised to us, since on this basis one could eliminate a good three dozen tourist excursions to

147

places where monuments are being restored.

During the eight-year existence of the excursion to Shakhmatovo the Bureau of Excursions has not received a single complaint about it (not about anything). On the contrary, all of the tourists are struck by the unusually poetic quality of this corner outside of Moscow, and by the unforgettable landscape, which in all likelihood remains the primary memorial treasure of these places. We can cite numerous expressions of gratitude from tourists, and precisely concerning the necessity and advisability of this excursion. In our opinion, the Bureau of Excursions should be called upon, considering the coming major anniversary of A. Blok, not to close but, on the contrary, to publicize this trip. In essence the complaint most of all concerns defects of a technical nature (poor means of transportation, fatigue on the road), for which the Bureau of Excursions, of course, assumes responsibility. But this is already another matter, a very serious one, but one not having any direct bearing on the question of the advisability of the given excursion. We feel that the excursion to Shakhmatovo is necessary, and that the technical defects that existed during the specific excursion under discussion do not provide the basis for a consideration of the question of canceling the excursion. This is the sole complaint about an excursion that has been operating for eight years.

We can smile at the naive words to the effect that "the only monument still in unsatisfactory condition is the church in the hamlet of Tarakanovo"; we can smile because overall this *is* the only "monument"—after all, there is nothing else there, if one doesn't count the stone.

Let us note that the following argument could be advanced only for the purpose of self-protection: "Not to show these territories on the basis of a church's being restored there seems ill-advised to us," since thus far the church is not being restored at all (and is not even close to being restored), but continues to exist in a state of ruin.

I will agree with some very important lines in the letter: "all of the tourists are struck by the unusually poetic quality of this corner outside of Moscow, and by the unforgettable landscape, which in all likelihood remains the primary memorial treasure of these places."

In fact, it would seem that there is nothing to show. The church at Tarakanovo is actually a pile of bricks. As for the photo exhibit—photographs of Blok, his mother, father, wife, aunts, autographs—one can see copies of all of this in editions of Blok at the library or even at home. At Shakhmatovo there's neither a dwelling nor suitable vegetation.

However, the landscape and paths (so long as excursionists walk along them); the very enchantment of the countryside, which seems illuminated by

Blok's poetry throughout the entire excursion; the narration about Blok by
trained and interested tour guides; the tour guides' reading of his poetry, which
(I should state plainly) not every tourist knows by heart—in short, I have in
mind a sojourn over the course of several hours in the atmosphere of Blok, and
the contact with him—all of these things in an amazing way satisfy those who
have come here, and they leave feeling enriched and more spiritually alive, as
if they had taken a drink from a clear spring. But if later on I were to discuss
the trip once more with the participants and ask: "Well, what in fact did you
see that was concrete, material, of museum quality?" Of course, essentially—
nothing!

And if they had visited the restored Shakhmatovo, with its house, picturesque garden, pond, annex, granary, jasmine, dog-rose bushes, well-swept
walks and flower beds in bloom?

And if the surrounding hills, villages, and Lutosnya River had been left
undisturbed? And if to all of this we were to add Boblovo (the restored Boblovo), without which it's impossible to imagine Greater Shakhmatovo, as we
agreed at the beginning of this essay to call all of these places Blok knew, taken
together?

Concerning my own acquaintance with Shakhmatovo, it took place on the day
of the first Blok celebration, August 9, 1970.

At that time I had no plans to write about Shakhmatovo, and viewed all of
this completely from the standpoint of festive curiosity. And I didn't have any
time for it, anyway. After we had walked from Tarakanovo, I knew that now I
would have to make a presentation along with the others. The multitude of
people who had gathered, and their strolling and sitting in the clearing (while
to the side of the clearing little groups gathered around unfolded newspapers
with simple eatables on them), dispelled the atmosphere in which I possibly
would have found myself if I had come here alone, and, if so, a lyrical and gray
day would have been better than the blinding radiance of these skies.

But I do remember a few things. In the ruins of the Tarakanovo church at
the time when we had walked up to them, to our amazement a film matinee
was in progress. Somehow, among these ruins a small film auditorium for collective farm workers had been set up out of boards. Right before our eyes the
members of the audience poured out of the ruins after the end of the matinee
onto the small green meadow in front of the church.

When we were walking through the village of Osinki, through a narrow
walkway between the houses and vegetable gardens, by some sort of intuition,
seeing an open back gate, I walked through it and, being well acquainted with
such peasant yards, cast a glance around the entire yard: I saw where the
sheepcote, the cowshed, and the chicken roost were, and the joiner's bench,

149

and the basket, and the pitchfork. . . . I entered only in order to see an authentic peasant yard (living in Moscow, one doesn't get to see one every day), but, by an improbable coincidence, I immediately saw on the wall, on a sturdy wooden hook on which at one time in the past a yoke undoubtedly had hung, the bronze frame of a piano. Desiring to learn how part of a piano had wound up in a crowded peasant yard, why it was there, and for how long it been hanging there, I entered the inner porch, where another surprise awaited me: not quite a trestle-bed, not quite a sofa, but, in general, a wooden stove-bench on four legs. Moreover, the two outer legs turned out to be turned, paunchy, and black, on little brass wheels, while the two remaining ones were made from ordinary wooden, squared beams.

The yard was open, but the house was locked, which often happens in villages. There was no one I could ask about the frame and the legs on the sofa. But on this day we were living and breathing Shakhmatovo, and, I really don't know whether it's accurate or not, but at the time I was convinced that these were the remains of Blok's demolished piano. After all, they didn't drag these parts of a musical instrument to Osinki all the way from Moscow. And we must bear in mind that Osinki is the little village closest to Shakhmatovo, a mere kilometer away, one only has to cross an oat field. It's also interesting that the following year, on a similar August day celebrating Blok, nothing remained in this yard anymore. Apparently the newly awakened interest in Blok had compelled the owners of relics to put them out of sight, out of harm's way. That's how it goes: this item had hung there for fifty years, sharing with Shakhmatovo itself the fate of oblivion, and it wasn't needed, and didn't ask for any bread, and suddenly everyone began talking about Shakhmatovo, Shakhmatovo, Shakhmatovo. Many cars began to arrive; there were thousands of people; some gave speeches and reminisced. No, it's better to put it out of harm's way.

. . . I asked Stanislav Lesnevsky to keep me company and make the trip with me to Shakhmatovo and Boblovo in order to take one more quick look around before I sat down to write my essay.

The first thing we did was stop at Solnechnogorsk, where, not far from the station in a wooden house containing many apartments, in a small, separate room, we located a withered, living old woman. That is, "living,"[77] not in the sense that the old woman was alive, but in the sense that liveliness shone in her movements and speech, and in her eyes and her entire character.

This time Ekaterina Evstigneevna Mozhaeva, because of her poor health, couldn't go with us to Shakhmatovo and take us around, show us where everything had been. But it turned out that Stanislav had already taken her there, and that, moreover, television crews had filmed the old woman against a background of the local bushes and trees. They had all become very interested in an episode she had related about the planting of a rose bush. As refracted through my perception, it went like this. Blok was planting roses. Katya, who at that time was a young woman if not an adolescent, and who in all likelihood was extremely pretty and fresh-skinned, was passing by not far away. The steward Nikolai, in order to give the poet an opportunity to admire this living flower, as a joke probably, asked Katya to help the gentleman plant a rose bush, to hold it while they filled in the roots with dirt. Katya carried out the request, but when they found themselves next to each other by the cutting that was being planted (and Blok himself, you know, was extraordinarily handsome), Katya realized that they had had some fun at her expense, that they hadn't needed her help there at all, and that two adults could easily have planted the rose bush without her. Then she blushed and, in her words today, "blurted out" something and "said something ridiculous" to them. It seemed that Blok was laughing gaily all the while.

Today Ekaterina Evstigneevna couldn't remember at all exactly what she had "blurted out," what she had said that was "ridiculous." But it was this very story about the planting of the rose bush that the television crew had filmed at Shakhmatovo. They began to ask Ekaterina Evstigneevna to recall the surroundings more precisely. It seems that the rose bush planted that day was intact and growing at Shakhmatovo to this day. Or at least in its place a descendant branch was growing that had sprung from its roots. It grows today among tall weeds and stinging-nettles, but Ekaterina Evstigneevna still found it.

"All right, Ekaterina Evstigneevna, that means you were standing right here, facing in this direction," the cameramen tried to establish, "and where was Blok standing?"

"Blok? Right over there, where that little tree is," the old woman pointed to a young arboreal sprout as high as a man. "Heavens! Isn't that him standing there? He really is standing there, after all!"

At this point it seems that a mystical trepidation came over the cameramen and Stanislav and, as they say, the blood froze in their veins. But it goes without saying that on the spot where Ekaterina Evstigneevna had imagined the image of Blok, they didn't see anything except the little tree.

Already questioned many times by various people during the preceding years, Ekaterina Evstigneevna couldn't relate anything new. Yes, she had taken mail to Shakhmatovo, had washed clothes, had washed dishes, and had been hired as a day laborer for various kinds of work. At that time they—the Mozhaev family—had lived in Osinki.

"Even now in the village," said Ekaterina Evstigneevna, "a peasant man has floorboards in his *izba* from the Shakhmatovo floor. Several pieces. I walked into the *izba* and saw the floorboards from Shakhmatovo."

"Maybe you were mistaken. Boards are boards. Why are you certain that the floorboards are the same ones?"

"I washed them so many times, how could I not recognize them!"

As old people sometimes repeat and lament to themselves something like "Lord, Lord, have mercy on me, Lord," so Ekaterina Evstigneevna would whisper from time to time, sighing deeply, moving her lips in words unrelated to those of our immediate conversation: "Shakhmatovo, my Shakhmatovo, my poor Shakhmatovo."

The Tarakanovo Highway led us from the flat area on which Solnechnogorsk stood (and nearby is the completely flat Lake Senezh), into hills, hills on both sides, into a landscape enclosed and bounded by the hills, and yet defined by them too. "The walls of Blok's study," I'll repeat after others the apt description of the Shakhmatovo environs tossed off by Andrei Bely.

In the hills are little villages and hamlets. In the past they had been distinctly identified by little white churches (let's recall that from a certain felicitous spot one could see twenty little churches); but now they are not so distinct from each other, and from a distance merge with copses, bushes, and the green earth.

"We have to ensure that the environs of Shakhmatovo are decreed a preserve and game reserve," said Stanislav, "it wouldn't be a bad thing if a small piece of the territory outside of Moscow were to retain its primordial beauty."

"What do you mean? You know you can't put up all of the churches again."

"But at least they should be prevented from constructing tall concrete buildings on these hills. Otherwise it will turn out like the hamlet of Novoe, where a sanatorium is housed in a former estate, and where they're planning to erect several multistory buildings. The whole enchantment of these places will be destroyed."

"Yes, that's true. This concerns not only the beauty of the object but also the beauty of its surrounding environment. The memorial object, to use the language of resolutions, decrees, and documents, can be ruined without even lifting a finger against it. Architects call this the 'moral destruction of a monument.' For example, an architectural structure as conceived by the architect-artist must be viewed against a background of green expanses and the sky. In front of it there also must be a vast, green, level space. And so, if an extended gray or brick building were to be erected behind it, the perception of this monument would be violated, its artistic effect would be weakened. But we can ruin the effect in an even simpler way: by erecting a long building not behind the architectural ensemble, but in front of it. Then it will be impossible to view this ensemble from a distance, but instead only up close, after one walks around the new building. And they could put up buildings on all sides of it. That would be the complete moral destruction of the monument, even though in a formal sense it had not been demolished. The next time you're driving in your car along the metrobridge from Lenin Hills in the direction of Komsomol Avenue and the center of Moscow, observe how precisely the delicate, pale-pink ensemble of the Novodevichii Monastery is blocked out by the gray blind of some building. When you enter Red Square from the side of the Historical Museum, you can't help noticing that the Cathedral of St. Basil the Blessed no longer appears against the background of the sky, as if floating or soaring in the air, so light, but against the background of the flat and gray blind of a new hotel, which by its very presence seems to increase the cathedral's weight and make it earthbound. Architecture is a delicate thing!"

But since we're concerned here not so much with architecture as with landscape and its protection, we'll turn to other obvious examples.

Take Arkhangelskoe outside of Moscow. A wonderful palace with a wonderful park surrounding it. There comes a moment when, after walking through the entire park (if you turn around, you can admire the palace from a narrowing perspective), you come out onto the farthest point of level ground. The earth falls away below on a steep slope: the park and palace are situated on the high bank of the Moscow River. Before your eyes, below and in the distance, extend the river and the expanses beyond it. To the joy that you have received from the beauty of the palace and the park's lanes, from the spacious, level lawn in front of the palace, from its cleanliness and orderliness, from the harmony between the architecture and nature, is added the joy of seeing a wonderful landscape that suddenly opens up before you. You have already prepared yourself to perceive something wonderful, and indeed, you are not deceived or offended by something ugly and deformed. No one spits unexpectedly into your soul, which has opened itself in order to perceive something wonderful. The clear mirror of the river, clear meadows on the other bank, and copses. A small haystack. No ditches, no protuberances, no garbage, no stripped earth, no wounds, no abrasions, no clutter, and no dissonant concrete bump of any kind. The grounds have been left the way they were when the palace was built and the park was laid out. Well, what can I say, everything is as it should be: the park and palace have been decreed a preserve, since this is an artistic and historical treasure, the property of our nation, as they say, a treasure that must be taken care of. They acted wisely when within the bounds of the preserve they included, besides the palace and park, the land that could be seen from the palace and park. If we were to describe a semicircle with a compass, we would have to move apart the legs of the compass by about three or four kilometers, not more. Herein is the entire field of view. For such a vast country as ours it's not a great loss. And moreover, it's not a loss at all, but on the contrary, the care and preservation of beauty. And, you

153

know, in the final analysis there may not be anything more valuable than beauty.

But let's allow the following scenario: you're walking through the park and come out onto a high vantage point, and there, on the other bank of the river, heaven knows what has piled up and accumulated. Oil storage tanks, a pile of concrete parts, a motor depot, smokestacks sending up black smoke, land crisscrossed by automobiles and tractors, spools of wire, the frames of some kind of cars, a fence made of reinforced concrete blocks. . . .

All of these things are necessary for the country's economy, but would we be acting in a proprietary manner if, in light of the country's huge expanses, we heaped up all of the aforementioned items (and many other things) before a landscaped, open square? After all, even an apartment contains various cupboards, pantries, attics, and secondary rooms for household needs, but there is also a living room, and in houses of the past there were attics and rooms for young women. I didn't dream up the picture containing the cluttered-up and disfigured countryside as viewed from a protected place. All of it looks just like this, if we no longer consider Arkhangelskoe (which has been preserved wisely), but the thousand-times-more-precious and wonderful ancient place known as the hamlet of Kolomenskoe.

The out-of-town residence of the Moscow sovereigns was located about seven versts from the city gates, on the high bank of the same Moscow River. The celebrated Cathedral of the Ascension, known throughout the enlightened world, and built by Vasily III out of gratitude for the birth of an heir, the future Ivan the Terrible. The enchanting, snow-white, blue-domed (topped with five cupolas) Kazan Church, built by this same heir, already grown, in honor of the conquest of Kazan. There was another wooden royal palace— the eighth wonder of the world—that alone embodied all of the characteristics of ancient wooden, large-scale tower architecture. Unfortunately, it did not survive, but was demolished during the reign of Catherine II. Either it would have been difficult to restore it in those days, or the empress, captivated by rococo and baroque design, could not fully appreciate the charm and uniqueness of this palace. But, incidentally, an exact model—down to the last little wooden log—has survived, and is kept today in the Kolomenskoe Museum, so that, with modern construction techniques and in light of our enlightened views on the ancient past, it wouldn't be any trouble to restore it. But at the moment we're discussing something else.

Having walked around and looked at everything there is at Kolomenskoe, we walk out (just as at Arkhangelskoe) onto the very high bank of the Moscow River. At this place the river forms a beautiful bend, while beyond it at one time extended level, green meadows. But there also exists a fundamental difference from Arkhangelskoe. There the palace stands in the depths of the park, and one has to walk rather far from it in order to come out onto the bank of the precipice, a slope; but here the main palace, the main architectural beauty, rests right above the slope, so that while one approaches the palace all the while one sees beyond it and together with it, in the form of an unavoidable background to this unique architectural treasure, all of the vast countryside outside of Moscow.

Now let's take a genuine, very beautiful piece of fine jewelry (it could be anything—a brooch, pendant, necklace, bracelet, or diadem), and let's lay it first on a piece of clean, smooth velvet, and then, for comparison, on potato peels, wood chips, iron filings, garbage that has collected in the corner under a broom, or even in a box containing household accessories, on a pile of buttons, spools, needles, thimbles, and pieces of cloth. Is there a difference or not? Similarly, Kolomenskoe, as something precious, "lies there" and is viewed against a background of all of the aforementioned items (and it's even worse than that), instead of being viewed on clean, smooth velvet. If you go there, you'll see for yourselves; don't let laziness keep you from making the trip to Arkhangelskoe in order to compare how it should and should not be done, how it should and should not appear.

I have in reserve still one more, the most scandalous, example of this type, linked with the most sacred place, the dearest name. Alexander Sergeevich Pushkin. Mikhailovskoe and Trigorskoe. A national preserve.

How can I describe all of this as simply and graphically as possible, especially for those who have never been to Mikhailovskoe?

Let's begin with the fact that Pushkin liked to stay at Trigorskoe, the house on the estate of his neighbors, the Vulfs. "Farewell, Trigorskoe, where I knew happiness so many times." He would go to Trigorskoe on horseback, but often went on foot as well, since he liked to take walks. The distance from Mikhailovskoe to Trigorskoe is about four kilometers.

This was his lyrical road, the path of emotional suffering, reflections, and emerging rhythms and lines of poetry. And this path was also beautiful. You see, Mikhailovskoe and Trigorskoe stand on hills, while between them is a bright-green valley adorned with the very winding Sorot River. Just imagine: you come out of the park at Mikhailovskoe and find yourself on a high, open hill overlooking a green valley, and you see that beyond this valley is another, even higher hill—Trigorskoe. Between the two hills is a path through clear, bright verdure. The river meanders at the bottom of the valley. The entire landscape seems to invite you to take a walk, to cover these four kilometers without hurrying; it predisposes you to ponder Pushkin's lyrical way of life. This valley between the hills was very likely the greatest treasure in the Pushkin preserve. You did, after all, walk and the entire time have before your eyes exactly the same landscapes as Pushkin did when he used to walk here. By means of this landscape you seemed to feel closer to him, felt a spiritual and emotional affinity with him, became his unwitting traveling companion, and experienced the same emotional suffering as he did. Unwittingly, you began to repeat lines from Pushkin's poetry, and isn't this because they possibly appeared for the first time on this path?

And so, what did they do there? Tourists, after all, have to drive up to the object of their observation in myriad buses. They have to drive up to a place from which Mikhailovskoe and Trigorskoe would be equidistant. If they were to bring the road, so to speak, to the backyards of these two hamlets in such a way that it would not be visible, they would then have to lay down two roads, and this would require an extra five kilometers of concrete roadbed. You understand, of course, dear reader, which decision the organizers of the preserve arrived at, the awful truth has already formed in your brain, although you refuse to believe it and hope that the reality nevertheless is not so terrible. But you are right. They carved out a concrete road right in the lowlands between Mikhailovskoe and Trigorskoe. But the road wasn't enough; in addition, they established several concrete squares as a parking lot for buses. Now, upon walking out to the edge of the Mikhailovskoe hill, you view, not the lyrical road of Pushkin, not a beautiful and clean valley, but something akin to a garage or motor depot. The valley does not invite you to take a walk, to stroll through it to Trigorskoe. On the contrary, everyone wants to turn away, to leave as quickly as possible in order not to look at the disfigured valley.

The widely known director of the Pushkin preserve, P. G. Geichenko, who really did a great deal to improve the places Pushkin knew and to popularize them, but who nevertheless allowed the negligence I describe here to take place, agreed with me. "Yes, we were in a hurry. The tourist wave swept over us. Well, that's all right, someday under calmer circumstances our descendants will set it right."

I confess that what amazed me most of all in this account was the possibility of such treatment of a preserve. Indeed, if a location is decreed a preserve—that is, must be preserved in its original state—how is it possible to ruin it with a concrete road and a parking lot? (In addition, they completely stripped the earth with bulldozers when they were filling in the roadbeds.) If it's really permissible to build a road and a parking lot, then why is this place called a "preserve"?

While making such observations (we even stopped from time to time, and I attempted to draw some schematic pictures of Mikhailovskoe for Stanislav, of how the road could have appeared there, and of where it leads today), we entered the Blok hills, which, one must hope, will also someday be legally made a preserve by the state, like the places Pushkin, Turgenev, Lermontov, Nekrasov, Tolstoi, Tyutchev, and Chekhov knew.

Here Stanislav knew all the "vantage points." Sometimes we would turn off of the Tarakanovo Highway and make a detour of a kilometer or two; we would ascend a hill, get out of our car, and from a high place survey the hills on the opposite side of the valley.

While standing on a high hill near the village of Sergeevo and gazing at the opposite high hill, where the hamlet of Novoe is located, we first regretted, it goes without saying, that this hill was no longer adorned by a small ancient church (we could easily see the little square on the slope of the hill where the church formerly had stood); and second, we distinctly imagined how the landscape would look if, on that spot, out from behind the trees, we were to erect four nine-story gray concrete buildings of a sanatorium—such a project exists. As if it would be impossible to construct several two- or three-story buildings, in which, incidentally, it would be much more comfortable to live than in high-rise towers.

At Shakhmatovo itself we didn't see anything new, but we hadn't planned on it, either. We went there without any special purpose, just to take another look at the remains of this lilac bush, of that dog-rose bush, and to walk along the path. Each of us in turn climbed up onto the stone and declaimed to the wide, empty, clearing: Stanislav—"Autumn Love" ("When in the foliage, damp and mildewed . . . "), and I—the introduction to the poem *Retribution*. We stood for a while on the spot where the house formerly had stood.

Incidentally, this is high ground, and meadowsweet, as is well known, grows in lowland areas, along ravines and the banks of streams, where it's dampest. This means that the meadowsweet we see at Shakhmatovo, among stinging-nettles, tall weeds, and bushes, is the meadowsweet planted at one time by Blok and even mentioned in his notebook. The meadowsweet and Sakhalin buckwheat remain intact, have survived, and vegetate among the present stinging-nettles; I didn't see a single mallow there, not to mention philadel-phus.

We drove to Boblovo. But at Boblovo, if you don't consider the village but only the Mendeleev estate, there's also nothing to see. The village still exists so far, but it, too, has been affected by changes. Some of the inhabitants have moved to the cities, and the houses have thinned out; the remaining houses have been altered somewhat: little terraces in the current style, slate on the roofs, television antennas. . . .

However, in their own time the Boblovo peasants really were lucky—it was their lot to reside in such a beautiful place! It's not for nothing that, as soon as Dmitry Ivanovich Mendeleev stepped onto this hill, he never again wanted to leave it. The height, breadth, distant expanses, haze of forests, blue of the skies, whiteness of the clouds, blooming grasses all around as high as a horse's belly, solitude, fragrant solitude—a royal place.

The Mendeleev house stood by itself away from the village, and no one could show us exactly where it had been. On an open spot at the edge of the forest, in the most visible (no matter in which direction one looks) place, we found stinging-nettles and holes from some kind of former structures, and it seemed possible that the house had stood here; however, for some reason the local residents kept referring us to the depths of the park, which has grown wild now. Yet it is well known that from the windows of the house there opened up expansive views not blocked by the trees. The principal charm of the estate lay precisely in this feature.

It is very strange that what has survived intact is the famous barn in which the household Mendeleev-Beketov-Blok performances took place, in which Ophelia was radiant in her cloak of flowing hair and holding an armful of flowers, while Hamlet declaimed his tragic monologue.

The barn is very long, low, sagging along its spine, covered with aspen shingles and blackened from the rain and winds by, it seems, the same blackness that with time covers antique silver. The state collective farm doesn't make use of this barn in any way now, but then inside it was neither clean nor swept out. In small piles on the earthen floor the grain that had been left here many years ago formed little mounds, at first becoming damp, then sprouting, and later becoming compressed until it was as hard as rock. Mold. Gloom. Emptiness. Neglect. A large bird glided noiselessly above us and flew out through an opening where there once had been gates. Only a second later did I realize that what had flown out of the barn was an owl, which had hidden itself here away from the daylight, and which in no way could have expected that people would once again appear within these walls. . . .[78]

"And so, Stanislav, let's discuss this point by point: what could be done with all of this, if you and I were gods? Let's formulate it."

"First: Establish a historical-cultural-environmental preserve (game reserve) for the purpose of protecting the landscapes that Alexander Blok glorified in his poems about the motherland: from the village of Vertlinskoe on both sides of the Tarakanovo Highway, in the areas surrounding Shakhmatovo, Tarakanovo, and Boblovo on both sides of the Lutosnya River right up to the Rogachyov Highway. Ensure that inside the boundaries of the preserve (game reserve) the cutting down of forests and the kind of construction that would deform the landscape are prohibited. Restore the dams on the Sestra and Lutosnya rivers. Preserve as an integral part of our historical memory and landscape the following villages in the environs of Shakhmatovo: Gudino, Osinki, Lukyanovo, Aladyino, Demyanovo, Tarakanovo, Bedovo, Dubrovki, Boblovo, Sergeevo, Fominskoe, Borodino, Zaovrazhie, Novoe, Merzlovo, and Kostyunino.

"Second: Restore the house of the Beketovs-Bloks at Shakhmatovo. Establish a museum-estate of Alexander Blok at Shakhmatovo as a branch of the State Literary Museum. Provide for the restoration at Shakhmatovo of the garden, auxiliary structures (the annex, *izba*, granary), and pond. Establish a protected zone within the bounds of two or three kilometers around the museum-estate that prohibits automobiles from driving up to the museum.

"Third: Restore the valuable eighteenth-century architectural monument, the former Church of the Archangel Michael at Tarakanovo, in which A. A. Blok and L. D. Mendeleeva were married.

"Fourth: Restore the estate and park at Boblovo as a memorial place of the great Russian scientist, Dmitry Ivanovich Mendeleev.

"Fifth: Declare the city of Rogachyov as a center and base for the tourist complex 'Shakhmatovo-Boblovo,' put in order the deteriorating cathedral in the center of town, and build a hotel for tourists.

"I think that's all."

I was amazed at the precision with which Stanislav formulated all of these points.

"Well of course! After all, I've recapitulated almost verbatim," laughed Stanislav, "the letter sent by the Writers' Union of the USSR to the Executive Committee of the Moscow Provincial Soviet. The letter was sent in 1975."

"And then what, did they agree with it, did anything happen?"

"I have an entire folder of correspondence with various establishments. In the majority of cases they're only formal replies."

Since Stanislav was always armed to the teeth, he opened his briefcase and rustled through some papers in it.

"Listen, from the curious to the serious: 'The residence of A. A. Blok at Shakhmatovo is marked at present by a memorial sign (a board); the directors of the State collective farm [!—V.S.] transported to Shakhmatovo a large block of natural stone, out of which a monument will be created [!!!—V.S.].'"

We both chuckled. Stanislav continued to leaf through his papers:

Resolution of the Executive Committee of the Moscow Provincial Soviet of May 31, 1977. The territory of the country estate of A. Blok, "Shakhmatovo," the former church, and building of the former country school of the hamlet of Tarakanovo have been put under governmental protection as a historical monument of local significance.

The Department of Institutes for Cultural Education informs you that the question of the restoration of memorial places linked with the life and work of A. A. Blok in the hamlet of Tarakanovo and at Shakhmatovo of Moscow Province, has more than once been examined by the Ministry of Culture of the R.S.F.S.R., as a result of which a special commission was created in order to study the complex of questions surrounding the perpetuation of the memory of the poet.

The commission's decree provides for: the establishment of a protected zone at Shakhmatovo, repair and restoration of the church, construction of a museum exposition devoted to the poet's creative work, and memorialization of the green plantations near Blok's estate.

The implementation of these efforts will require between 1.5 and 2 million rubles and a four-year period for their realization, with completion by 1980.

"But wait a minute, wait. . . . You know, two years have already passed."

"That's just it. Two years have passed. The year 1980 is almost upon us, and nothing has been done yet, they're waiting for—how did they put it?—'the realization.'"

"Maybe they don't want to spend the money."

"How much money are we talking about? And besides, the expenditures would quickly be recouped. Objects having to do with tourism pay for themselves very quickly."

"It's all the same: your entire correspondence won't accomplish anything, even if it fills more than ten briefcases."

"But what will, then?"

"As soon as one enters the main building of the museum at Karabikha, one reads the following inscription on the wall in large letters: 'By a resolution of the Soviet of Ministers of the USSR. For the purpose of the perpetuation of the memory of the outstanding Russian poet N. A. Nekrasov, the Soviet of Ministers resolves: to establish in Karabikha the museum-estate of N. A. Nekrasov, to entrust to the Yaroslav Provincial Executive Committee. . . . ' And so on. But institutes of cultural education of regional or provincial scope . . . do they really have the power to pass such a resolution? No, only the Soviet of Ministers does, then the wheels will begin to turn right away. . . . "

For a time we drove along in silence.

"Do you feel sad for some reason?" Stanislav finally noticed my mood. "Or don't you believe that it's possible?"

"What, specifically?"

"For a resolution to be passed, declaring Shakhmatovo and its environs a state preserve and restoring the church at Tarakanovo and restoring the houses at Shakhmatovo and Boblovo. . . . Is it really impossible?"

"It's indeed possible. And sooner or later it will happen."

"Then what, in your opinion, is impossible?"

"How shall I put it. . . . Well . . . it's impossible for a carriage carrying guests to drive up to the door from the direction of the Podsolnechnaya Station, and for Alexandra Andreevna and Maria Andreevna to meet the guests and take them into the living room, and for the rooms (in the words of the guest Andrei Bely) to make an impression of coziness and brightness. . . .

"And for them to step out onto the terrace and see Alexander Alexandrovich and Lyubov Dmitrievna returning from a walk. . . .

"And for the image of them 'to be impressed in relief: on a sunny day, among flowers.'

"And for: '. . . I looked out the window over the tops of trees near a corner of the garden where the land sloped downhill, at the horizon of an already pale-blue sky with slightly golden, ash-gray clouds. . . .'

"And for: '. . . I remember, that evening we walked from the house down the road that crossed the clearing, through the grove from which the ravine opened up; behind it was high ground and above it was a pink, a tender pink sunset. . . . A. A. said to me, as he stretched out his hand, "Over there is Boblovo." "I used to live there," said L.D., pointing at the sky. . . .'"

And the earthly sounds in all likelihood began to grow quiet in this early-evening hour, and already the grass under our feet had begun to grow cold and damp, and, illuminated by the sky at sunset, around these young, fine-looking, proud, sensitive, and intelligent people filled with a sense of their own dignity, "the Russian earth shone in its tender beauty."[79]

A TIME TO GATHER STONES

EVERY OBJECT HAS its o wn history, regardless of whether it is a cup of some sort, a walking stick, a book, or a mirror. It was bought in a specific place. First the object belonged to one person, then to another. It matters who these people were. It is one thing if a stick belongs to our boys, who use it as a bat in a game of *lapta*,[1] but another thing entirely if the stick once belonged to Anton Pavlovich Chekhov, who loved to walk with it in the Crimean Mountains.

And then the object was given to someone as a gift, or sold, or handed down through the generations. It was loved, it was beautiful, and it was useful. People drank from the cup, walked with the stick, and read the book, while the mirror reflected in itself the images of various people over the course of two or three hundred years.

Then the object may have lain about somewhere and been neglected, or it may have been lost or stolen. Someone found it, acquired it anew, cleaned the dirt from it, and washed it. The cup, for example, had been broken, but now has been glued back together. It is no longer possible to drink from it, but it is treasured as a memento and work of art, since it is an exemplar of antique china: Gardener, Sèvres, Blue Swords, Kuznetsov.[2]

The walking stick would stand in a museum. And in fact in the Chekhov House in Yalta there are sticks that were used by Chekhov on his walks. Who could possibly raise a hand to chop up such a stick, say, to use as kindling for a samovar? And yet it is not made of gold or porcelain but is just a piece of wood. Or if the staff of the philosopher and poet Grigory Skovoroda[3] were to be discovered now, wouldn't it be put in a museum, even in a general museum of literature (since I suspect that a museum devoted to Grigory Skovoroda does not exist)?

The mirror would be in a museum, too, especially if it were known for certain that the following had seen themselves in it: Pushkin, Zhukovsky, Gogol, the Kireevsky brothers, Alexei Konstantinovich Tolstoi, Dostoevsky, Lev Tolstoi, and finally. . . . Now tell me, would it be conscionable (at a particular level of civilized behavior) to throw away such a mirror? Or such a cup, if all the persons I mentioned above had drunk from it? Are you out of your mind— throw it away?! Put it under a bell jar right away! On velvet! In a special case! Blow away the specks of dust on it. . . . Attach a special warning system to it to prevent theft. . . . It's not a laughing matter if Gogol, Dostoevsky, and Tolstoi were involved! After all, this is our national treasure. And not only ours, but a treasure of the entire, as they say, cultured and progressive human race. Irreplaceable cultural and spiritual treasures. For this reason, I think, such a cup or such a mirror even in material terms would be worth an enormous amount of money. Or, at any rate, worth quite a bit more than face value. If, of course, it could be proven without a doubt that, yes, all of these people had drunk from, had viewed themselves in, had touched. . . .

Now the Americans (who are rolling in money) are buying up and collecting everything. They are even buying the old bridges across the Thames from the English. At one time it was rumored that they had inquired about the price of the Cathedral of St. Basil the Blessed. Let's imagine how much they could pay (at an auction) for a Russian monastery built in the fourteenth century, with all its walls, towers, cathedrals, ancient icons, graves of well-known persons, bells, and a library containing thirty thousand volumes (plus ancient manuscripts numbering in the hundreds of thousands); and, in addition to this, if this monastery were linked with the memory of such Russian writers as Gogol, Zhukovsky, A. K. Tolstoi, Turgenev, Dostoevsky, the Kireevsky brothers, Zhemchuzhnikov, Apukhtin, Maximóvich, Solovyov, Leontiev, Lev Tolstoi. . . . If it were known for certain that Tolstoi, having left Yasnaya Polyana,[4] had stayed first in this very monastery, and that Dostoevsky had based his best novel, *The Brothers Karamazov*, on this very same monastery?

"Where is such a monastery being sold?" the buyers of treasures would cry. "Wrap it up, we'll pay top dollar!"

"No!" One could say to the Americans. "This is our national, cultural, historical treasure. It is not for sale; it is priceless."

Let's also not forget the fact that the object could be inherently beautiful and valuable (at face value) even if famous people had not touched it. But if both factors were present, there would be nothing to discuss. Put it under glass and on velvet!

Let's take a look at this object, this cup, this mirror, this book (whichever object most appeals to you), from which people drank, in which they looked at themselves, which all those persons mentioned above had read, or at least touched (not including the hundreds of thousands, or perhaps even millions, of ordinary people); let's recall two or three episodes from the history of this object, assess its current condition, and ponder its future fate.

This object is called "Optina Monastery," and it lies in a rather neglected state three kilometers from the city of Kozelsk in Kaluga Province, on the banks of the Zhizdra River.

I don't remember now which writer (was it Maxim Gorky?) described a Russian peasant who sat down at a fork in the road and grew thoughtful.

"What are you thinking about?" someone asked him.

"Well, I just can't decide: either I go to the monastery or I become a robber."

It's difficult to say whether monks often ran away from monasteries and became robbers or whether this sort of thing happened at all, but the opposite phenomenon undoubtedly was not uncommon. At least in oral folk art—in songs and legends—this is a very popular subject. Let's take even the well-known song "There Once Lived Twelve Robbers, There Once Lived the Chieftain Kudeyar," performed so superbly by Fyodor Ivanovich Shalyapin, with choral accompaniment. Or take the song "On the Old Kaluga Road at the Forty-Ninth Verst." Or let's remember Uncle Vlas in Nekrasov's poem.[5] First they all brandish their bludgeons and spill the blood of "venerable Christians," and then they repent and enter a monastery in order to pray for the forgiveness of their sins. To be sure, Uncle Vlas chose a different path for salvation: he wandered through Rus and collected half-kopeck pieces for the construction of churches.

> He walks in the coldest winter,
> Walks in the summer heat,
> Asking all of Christian Rus
> To give whatever it can, . . .
>
> And they give, they give, the passersby . . .
> And out of all their efforts
> The cathedrals of God rise up
> Across the Motherland's face. . . .

Thus the robber known as "Opta," who lived sometime in the fourteenth or fifteenth century, having brandished his bludgeon enough (at that time one could brandish one's weapons in complete freedom—the forests were dense and there were few travelers, while communication between populated areas did not exist), finally repented and decided to save his soul. In a remote pine forest on the banks of the Zhizdra River he established a monastery, which at that time probably consisted of a few mud huts for five or six people and a little chapel for prayer.

At that time the city of Kozelsk already existed and had even become famous for its heroic defense against the hordes of Khan Batu.[6] The city fought for seven weeks and consequently was massacred down to the last inhabitant. The young prince of Kozelsk was drowned in blood.

However, by the seventeenth century the city once again was flourishing, with a population of five thousand, forty churches, and a stockade.[7]

In 1776 Kozelsk was awarded a coat of arms, and the decree of Empress Catherine the Great on this matter stated that the citizens of Kozelsk " . . . by their very death had demonstrated their loyalty. In memory of this event[8] they deserve a coat of arms: against a crimson background, symbolizing bloodshed, are arranged horizontally five silver shields with black crosses that express the courage of their defense and their unhappy fate, and four gold crosses that indicate their loyalty."[9]

I mention Kozelsk only because its proximity to the monastery (three versts away) makes one have doubts about the romantic legend of the robber Opta. A well-known robber would hardly have settled in an area so close to a city, where both people and the authorities lived. The city of Kozelsk in and of itself is not linked with the founding of the cloister and merits attention here only as an orientation point indicating its location. On the contrary, historical sources point to the ongoing conflicts between the city's inhabitants and the monks who settled on the opposite bank of the river.

The crux of the matter was that during the reign of Boris Godunov,[10] to pray for Tsar Fyodor Ioannovich, who died in 1598, the monastery was given, "for candles and incense . . . a place for a windmill, and on the bank opposite the windmill an enclosed place for another windmill [in order to build a barn— V.S.], and in addition, fishing rights to be shared with the farmers."

The windmill and fishing spot near it (all of this is on the little river named "Drugusna," which at this point flows into the Zhizdra) were the main source of income and the sole means of food for the brotherhood. And everything would have been fine: they would have continued to mill ordinary flour, accepting the flour itself as compensation, and they would have continued to catch fish, but there were hooligans in Kozelsk who evidently had no special liking for the monks and wanted to spoil things for them. They incited someone, and consequently "a certain Mishka Kostrikin built on the little river 'Drugusna,' without a tsarist decree, the only windmill approved by the Kozelsk dragoons and archers; in the process, he stopped the ancient windmill that stood upstream and destroyed it utterly."

In other words, the dam farther downstream blocked the little river, raised its water level, and damaged the monastery's windmill. Elder Isidor, who was the superior of the brotherhood (which at that time consisted of several hermits), sent a petition to Tsar Alexei Mikhailovich: "Optina Monastery has no peasant farmsteads[11] and material help is not forthcoming; they feed themselves through donations and their own labor. . . . I ask you to order that the windmill of Mishka Kostrikin be given to the monks in the monastery for candles and incense and wine for the church, and that it be given to the organizers of the brotherhood for their subsistence, so that they do not perish from hunger."

Alexei Mikhailovich died while the petition was on its way, and the ruined—or more precisely, blocked—monastery windmill stood silent and did not operate. In the meantime, Mishka's dragoon friends, out of revenge "having conspired with the leaders of the city of Kozelsk"—that is, having enlisted the support of influential citizens of Kozelsk—appealed to the still young sovereign, Fyodor Alexeevich, to allow them to keep Mishka's windmill for *obrok*.[12] However, Isidor and the brotherhood were also not simply biding their time: they sent their petition on July 13, 1676.

Having considered the matter, Tsar Fyodor Alexeevich sent a reply on the third of August (rather quickly, one might note, for only twenty days had passed) to the Kozelsk governor, Vasily Ivanovich Koshelyov, whom he ordered "to demolish Mishka Kostrikin's windmill and not to build any more windmills downstream from the monastery's windmill."

But the dragoons did not desist. Some time later, already during the reign of Peter,[13] two dragoons named Grigory Novikov and Nikifor Strygin, along with their friends, seized by force the hayfields that the monks had cleared with their own hands.

In response to the monks' latest petition, Peter wrote the following brief resolution, in keeping with his character: "Bring them in for questioning." Evidently the interrogation proved to go in the monks' favor, and subsequently all of their disagreements with the restless dragoons ceased.

The remainder of the monastery's history is of relatively little interest. The abbots changed, minor contributions were made to it by the neighboring boyars and landowners, and the construction of the first stone church was begun. To be sure, when you read the historical description of the hermitage, several sentences can rouse the imagination, if it has been prepared for this. For example, in 1684, " . . . the courtiers[14] Andrei and Ivan Shepelyov kept their promise and delivered the following holy icons to the monastery: the Icon of the Savior, the Icon of the Entrance of the Holy Virgin into the Temple, the Icon of St. Pafnutius of Borovsk, the Miracle-Worker, and the Royal Doors. . . . The Icon of George the Martyr 'in olden times, had been transferred from the village of Mormyzhev.'"

Quotation marks inside quotation marks appear at this point, since the entire extract was taken from the "Historical Description of the Kozelsk Optina Monastery of the Entrance of the Holy Virgin into the Temple" by Leonid Kavelin,[15] who in turn took the sentence concerning the fate of the Icon of George the Martyr out of a deposited monasterial book.

But it is impossible for the imagination not to be roused here, impossible not to wonder where all of these icons and the Royal Doors are now, particularly the Icon of George, if in the seventeenth century people were already saying that "in olden times [the icon] had been transferred . . . from the village of Mormyzhev."

It would be good if one of the local inhabitants had carried it off when the monastery was liquidated, but that is probably not the case. If only one could catch a glimpse of it, even out of the corner of an eye!

And here is a description of another object:

> An eight-bar silver cross. . . . This cross was donated to the cloister in 1850 by the Likhvinsk landowner Afanasy Nikolaevich Mikhailov, who inherited it from his relatives, the Kamynins, as the inscription engraved on the lower part of the cross attests: "Prayer of Grigory Ivanov, son of Kamynin." Popular belief indicates that the aforementioned Grigory Kamynin wore this cross during the Battle of Kulikovo Field, also known as the Bloody Battle of Mamaev. . . . A cursory examination of the cross suffices to establish the ancient workmanship on it, and judging by this, one can conclude that it belongs to a time in our country's history not yet conducive to the development of the arts, a time when of necessity people paid little attention to the external decoration of objects and took into account only the object's utilitarian purpose. Today the cross is kept in the sacristy.

Where is this sacristy located? Where is the cross? The book was published in 1885. This means that as recently as one hundred years ago one could take into one's hands this object, which had been on the Kulikovo Field. We mentioned earlier that each object possesses its own history, and here is confirmation of that fact. Is this not a splendid history? Only the final chapter remains unknown. The thread crossing many centuries has broken; it has broken and disappeared into the darkness.

Let's take yet another object:

> An enameled icon that was kept behind the abbot's place in the church; although it is not ancient, the monasterial craftsmanship on it is first-rate. It consists of thirteen separate icons: the central icon depicts the Resurrection of Christ, and it

is encircled by representations of the Twelve Great Feast Days. . . . It was donated to the cloister by the Muscovite merchant Vasily Ivanov Putilov. . . . One of
the best Rostov masters worked on this icon for an entire year and died soon after
finishing it. In our opinion, as one of the finest examples of the enamel artistry
that flourishes to this day in Rostov, it merits the serious attention of experts. A
similar icon on the spot cost approximately two hundred silver rubles.

Well, we won't speculate on how valuable in all senses of the word this icon
would be today. The experts will understand.

At times the monastery fell into ruin; at times it came to life somewhat; at
times it was abolished entirely and attached to another, larger monastery; and
at times it became active once again. The cloister at times dwindled to two
monks (and even so, one of them was blind). This little flame kept smolder-
ing, kept flickering in the wind of time, threatening to go out—and suddenly,
as if from nothing, it grew into a huge and bright flame. The shabby cloister
consisting of just a few wretched monks suddenly was transformed into a
wealthy and beautiful monastery (according to a traveler, "like a basket of
white lilies against the background of a deep blue forest") that was known
throughout Russia.

Well, not really "as if from nothing": it means that oil was poured on the little flame and kindling was put under it. And yet one can consider the flowering of Optina Monastery to be unexpected and sudden.

The first stimulus came from the Moscow Metropolitan, who turned his
attention to Optina and decided to support it. He found a practical, energetic
person (Abramius) and sent him to the monastery to carry out his wishes; most
likely he gave it financial support too, at first. Later, of course, things took
care of themselves. . . . Rus was devout and densely populated. A kopeck here,
a kopeck there, at first raining in drops, then a steady rain, and finally streams
of money poured in from all directions. When the monastery had grown
financially strong and acquired some splendor (several centuries of existence at
this location had already passed, and hence a tradition had developed), sizable
donations from wealthy individuals became possible, as well as wills leaving
inheritances to the monastery in exchange for prayers for the souls of the
deceased. As each of us is a sinner, it is somehow calming to be able to look
after our own soul by ensuring its remembrance and collective monastic prayer
for it for several hundred years. Thus we can explain the material growth of
any monastery of that time, whereas the reasons for a monastery's special popularity require some elucidation.

In many monasteries this popularity developed in tandem with the note-
worthiness of certain colorful individuals who lived there at various times and
who were popular with the common folk; these individuals subsequently were

proclaimed "miracle-workers" and elevated to the rank of sainthood. The following fit this description: Sergius Radonezhsky, who founded the Trinity Monastery of St. Sergius; Zosima and Sabbatius, the Solovetsk miracle-workers; Cyril Belozersky; Nil Sorsky; Pafnutius Borovsky; Joseph Volokolamsky; Seraphim Sarovsky. . . .

In times past pilgrims would say to one another: "I'm going to the Reverend Sergius," "I'm going to Tikhon Zadonsky," or "I'm going to Boris and Gleb." The popularity of the saints largely determined the popularity of the monastery itself.

But it's not at issue here that Optina Monastery, having vegetated for several centuries on the bank of the Zhizdra and having been embroiled in conflicts with the Kozelsk dragoons and with Mishka Kostrikin, subsequently did not acquire its own saints, miracles, and relics. In this sense the monastery did not develop a tradition, but instead its popularity was created suddenly in a rather peculiar way.

Even to explain in a word or two, I have to start from way back.

In a conversation with me, Nadezhda Pavlóvich, who in the 1920s had spent several years at Optina on a specific assignment (she had been sent there as a museum worker) immediately before its closing and liquidation, summarized the following precise scheme:

> Without Paisius Velichkovsky it is impossible to understand the Optina eldership; without the elders it is impossible to understand the Kireevsky brothers and the Optina publishing house; without the Kireevsky brothers there would be no Slavophilism as a philosophical movement of the nineteenth century. Without the elders there would be no connection between the monastery and Gogol, Dostoevsky, and Tolstoi. This means that we have to begin with Paisius Velichkovsky.

I don't intend to relate the entire story in detail, but it is necessary to outline a few facts.

Paisius Velichkovsky lived in the eighteenth century. He was born in 1722 and died in 1794. Born in Poltava (his given name was Peter) and being very religious, he traveled around to various monasteries for a long time until he reached Mount Athos. During his wanderings he realized that Orthodox monasticism was degenerating into external ritualism at the expense of genuine and profound (from his point of view) spirituality. While studying the writings of the so-called church fathers on Mount Athos,[16] Paisius noticed that Slavonic translations were often inaccurate and incomprehensible. For this reason he began to collect ancient manuscripts and to correct later copies against them. He began to search for the original sources—i.e., Greek texts—

so that he could translate them into Church Slavonic. For example, he translated the famous "Philokaliya," a collection of ancient patristic writings now known by the title "Love of Goodness." Nadezhda Pavlóvich has saved a copy of this book that contains the extensive marginalia of Alexander Blok.

Church historians agree that Paisius was not an innovator of any kind and did not have anything new to say, but rather he only tried to renew the traditions of the ancient church, purifying it of later features. He came up with the idea of starting a school for the instruction of Greek to young monks, who would then become able translators of ancient church literature.

Paisius finally settled down at Nyamets Monastery (on the territory of Romania), at which there numbered seven hundred monks. Here Paisius's literary activities developed on a large scale. An entire school of translators labored over the correction and translation of Greek and Latin books. Little by little, Nyamets Monastery became "the center and light of Russian Orthodox monasticism, the school of ascetic life and of spiritual enlightenment."

Moreover, Paisius corresponded extensively with other Russian monasteries. His students and followers, who were scattered throughout Rus, subsequently became the elders and abbots of many Russian monasteries, who embodied the ideas and concepts of their teacher Paisius. During the course of an entire century, right up to 1917, Russian monasteries felt the influence of Paisius Velichkovsky and his so-called (for all that!) ideas of revival, which were called a "spiritual renaissance" in pertinent church literature.

And thus Optina Monastery by the will of fate became the center of this spiritual renaissance. Speaking in contemporary terms, I'll note that Optina Monastery gradually became the nesting site for students and followers of Paisius Velichkovsky, and later for the students of his students, that is, for a second generation of his followers. Hence the monastery's own publishing house; hence the enormous carefully selected library holdings (mainly on monasticism); hence, too, the famous Optina elders who eventually attracted to the monastery the attention, not only of the vast, devout masses, but also of the most enlightened individuals of that time.

I don't know if there was some kind of singlemindedness of purpose in all of this, but nevertheless one suspects a certain system in the course of events. Starting from the year 1800, Paisius's student Theophanes had been living at Optina Monastery. It is possible that he influenced Philaret, the head of the Kaluga diocese, in his efforts to establish a skete[17] at Optina, and to this end he summoned two more of Paisius's students—Moses and Anthony—from Smolensk Province. As soon as the skete was ready, yet another student and friend of Paisius—Hieromonk Leonid—moved there from the island of Valaam. The famous Optina line of elders begins with him.

During the course of the nineteenth century there were three elders. Each replaced the next one as if by right of succession; moreover, each new elder lived in the skete for several years with his predecessor, under his tutelage. Such was their continuity. Elder Leonid died in 1841. He was replaced by Elder Macarius ("the Gogolian"), who lived until 1860. From this date through 1891, the most famous of the Optina elders—Ambrose ("the Dostoevskian" and "Tolstoian" elder)—lived in the skete.

Elder Leonid came from the merchant class; Macarius was descended from the nobility; Ambrose was born into the family of a village priest in Tambov Province. Leonid's actual surname was Nagolkin, Macarius's name in secular life was Mikhail Ivanov, while Ambrose was named Alexander Grenkov.

At this point one might ask what kind of phenomenon these elders repre-sented and what eldership in general signified in former Russian monasteries. But Fyodor Mikhailovich Dostoevsky posed precisely this question in a chap-ter of his novel,[18] appropriately entitled "Elders." Dostoevsky discusses the essence of the matter in great detail (and of course every contemporary reader has a copy of the novel at hand), but still we can cite several lines from it that provide the key to understanding the popularity of the Optina elders.

Many said about Elder Zosima [read "Ambrose"—V.S.] that, in allowing for so many years all those who came to him to confess what was in their hearts and who yearned for his counsel and healing words, he admitted into his soul so many revelations, contritions, and confessions that he finally developed very keen insight: upon first glancing into the face of a stranger who had come to see him, he could guess why the stranger had come, what he needed, and even what kind of torment was troubling his conscience—and he would astonish, embarrass, and sometimes almost frighten the supplicant by such precise knowledge of the lat-ter's secret, even before he had uttered a word.

Moreover, Alyosha almost always noted that many people, almost everyone who came to see the elder for the first time for a private audience, entered in fear and agitation but almost always left in a radiant and joyful state; even the gloomi-est face was transformed into a happy one.

One must imagine a Russia of that time in all the complexity of her social relations and daily life—peasant, merchant, landed gentry, bureaucrat, cler-ic—and the mutual interaction of all these spheres of existence: a Russia with hundreds of thousands of churches, with thousands of monasteries, fairs, patron saints' days, marriage ceremonies, and funerals; a Russia with a multi-tude of families, whose masses were almost illiterate (in any case, television and radio did not exist at that time, nor did the magazine *Zdorovie* [Health]); but a Russia yearning, first, to express herself, and, second, to hear words of admonishment, instruction, and reassurance. This craving to express what was painful and to hear words that would heal on one level produced and infused spirituality into all forms of nineteenth-century art, while on another level it gave rise to streams of semiliterate or completely illiterate believers, semibelievers, agnostics, those wavering in their belief, and seekers—all making pilgrimages to the elders grown wise with experience.

Of course, every village had its own priest to whom one could go for confession and advice. After all, faith was faith, but it was essential that the priest be a personality, too, and was every priest one? In this matter there could be many various nuances. Moreover, "a prophet is not a prophet in his own homeland,"[19] one would not confess everything to Father Alexander, with whom one was going to live for many more years right here, in this very village. This also bespeaks the fact that the official clergy apparently did not always enjoy the requisite authority with its flock. Confessions took place and were even required, but what sort of confessions were they?

"Father, I am a sinner."

"Go in peace, and sin no more."

A formality. But an elder is a celebrity and, for all intents and purposes, holy. Stories about him are heard all over Russia, and you see him once in your life (as he, you), which means that you can confess everything to him up to the last detail, all your most cherished secrets; you can bare your soul to him. Your faith in his wisdom and perspicacity is boundless:

> Upon visiting the skete, Alexei Tolstoi saw the passion even in the asymmetrically placed eyes of Macarius, as the latter—wizened and completely gray-haired, with a crutch in one hand and a rosary in the other, to the general whispering of "the elder is coming"—appeared before the people and, blessing all those to the left and right, proceeded to his "shanty," sat down on a couch, and began to answer questions. He was accompanied by Ambrose (the secretary and correspondent who had been tonsured only eight years earlier), who wore a *kamelaukion*[20] and cassock, his eyes twinkling intelligently. Ambrose was enormously learned—he was a poet and theologican, knew five languages, and was designated as the successor of Macarius, which took place in 1860. At this time Ambrose began to acquire the knowledge that would make a lasting impression on Dostoevsky and Lev Tolstoi.[21]

The years would pass, and suddenly Ambrose would be—an elder.

> Downstairs by the wooden porch attached to the external wall of the fence, there was on this day a crowd consisting of women only, about twenty in all. They had been told that the elder would come out at last, and they crowded together in anticipation. . . . The crowd pressed forward toward the porch. . . . The elder

173

went up to the top step and began to bless the women crowding toward him. . . .
Many of the women in the crowd broke down in tears of tender emotion and
ecstasy produced by the intensity of the moment; others rushed to kiss even the
hem of his clothing; still others began to lament something incomprehensible.
He gave his blessing to all of them, and conversed with some.[22]

The official church was jealous. To be sure, of course, one cannot say that
these elders were not ecclesiastically affiliated people or were in conflict with
the church. On the contrary, they were Orthodox Christian elders, and not
some kind of heretics, although some official church hierarchies at times
attempted to identify them not only as heretics but even as Freemasons.

And why shouldn't they be jealous?! There was a monastery, an abbot, and
several hundred monks, but all this seemed only an entourage, the frame of the
setting in which lived—the elder. The monastery's fundamental authority
depended on the elder. The *hegumen* was virtually superfluous, or rather was
needed only as an intermediary. Everyone who came to the monastery,
whether on wheels or on foot, came because of the elder; everyone wanted to
see him, speak with him, and receive his blessing. The monastery's hostels
were full, shelters for the needy were packed to bursting, you couldn't force
your way into the monastery's churches—and all because of the elder. There
was something in this phenomenon that slightly bypassed the church itself, as
well as its priests, archimandrites, and metropolitans. People bypassed all this
to appeal directly to God, while the intermediary took the form, not of the
entire church with its complex organization, but of only a single, wizened old
man who nevertheless lived on the premises of the monastery, in the
monastery's skete. It is worthwhile just to imagine him living, not on the
monastery's premises, but simply in the woods, in a little hut, in some kind of
dugout, or in a cabin—the crowds of people would keep coming to see him all
the same, and in light of this, the reason for the church's jealousy and dissatisfaction, which often took the form of open persecution, becomes even clearer.
Elder Leonid, for example, was forbidden to receive visitors, but the visitors
kept coming: they waited for days on end, staying overnight and living near
the skete. The elder was transferred from the skete to somewhere inside the
monastery's walls, but the people still went there in droves. A high wooden
fence was built around the elder's cell and the gates were kept locked—it was
all in vain. Those longing to confess and receive counsel broke through all the
barriers, and finally the church had to give in: Leonid was allowed to return to
his skete.

While Elder Leonid exerted an influence mainly on the lower strata of faith-
ful and devout Russia, on simple monks and peasants, the second Optina elder,
Macarius, attracted people of the cultured strata of Russian society. His

enormous erudition, vast knowledge of books (not without reason were these two students of Paisius), noble descent, worldly experience, and knowledge of languages—all this made him able to converse on the same level as his visitors, be they the Kireevsky brothers, Gogol, or Alexei Tolstoi.

Of course, sometimes external circumstances would play a role. If, say, the Kireevskys' estate had been located somewhere near Pskov, perhaps such close ties between them and Optina Monastery would not have arisen; but the Kireevskys lived right nearby. Their estate, Dolbino, was located some forty versts from Kozelsk, a secondary though important consideration, since it wouldn't have been difficult for them to travel to Valaam and Solovki when they needed to, but, still, it would have been incomparably more difficult for them to develop close ties with such distant monasteries.

Russian public opinion concerning the Kireevsky brothers was not unanimous, as public opinion itself was not monolithic but was split into circles, directions, movements, and tendencies.

Concerning the younger brother, Peter Vasilievich, this was a relatively uncomplicated matter. A historian, he considered the collecting of folklore, of folk songs, to be the most important activity of his life. In the course of twenty-five years he unceasingly, with a genuine collector's enthusiasm, searched for, pleaded for, listened to, wrote down, and edited Russian folk songs. But this was not pure and, so to speak, mindless collecting in the manner of collecting stamps, when it doesn't matter where the stamp comes from (even from Australia, for that matter) so long as it is unique. Peter Vasilievich did not collect just any songs: these were Russian songs, his native culture, the soul of his people. Each year in summer he spent time in Dolbino and at Optina. He traveled on foot to various hamlets and villages in search of the pearls of folk poetry. In Optina's shelter for the needy he would listen for days on end to wanderers, pilgrims, and travelers from the forests of Archangel and Vologda, from the Ukraine, from Irtysh, from the Don Cossack region, from the expanses of Ryazan, from the little villages near Tver, and from the hamlets of Yaroslavl and Vladimir. News of his activities spread to Moscow and St. Petersburg. Countless volunteers helped him by sending him folk songs they had collected. Among these volunteers were such people as Gogol, Koltsov, and Dal. Even Pushkin sent Kireevsky a notebook of songs collected at Holy Mountain Monastery of Pskov Province. Approximately five hundred songs arrived from Belorussia. In the Optina library a special corner or, as we would say today, a department or collection was designated for Peter Vasilievich Kireevsky and his folk songs. To be sure, Elder Macarius constantly tried to direct Peter Vasilievich's activities to the collecting of ecclesiastical songs: and Peter collected these, too, but not exclusively and not primarily.

The songs he found were reworked and classified according to strict rules:

where the song had originated, when, and for what purpose (wedding, ceremonial, romantic), what type it was (solo, for singing and dancing in a circle),
and how it was performed. Peter Vasilievich was an accomplished musician
himself and wrote down not only the words but the melody, too. His only
inconvenience lay in the fact that in the library at Optina Monastery he was
not allowed to play the piano in order to check the notes he had recorded. To
this end he had to rent a quiet, secluded wing of the merchant Demidov's
house in Kozelsk.

Peter Vasilievich Kireevsky collected over seven thousand songs. If this did
not constitute a heroic deed, then it was selfless devotion to a cause, as the
poet Yazykov affirmed in the lines "In his devotion to his heritage he was an
enlightened ascetic."[23]

All this—devotion to his heritage, dedication to a cause, and enlightenment—to a certain degree also applies to the older brother, Ivan Vasilievich
Kireevsky, although this figure is somewhat more complicated. But nevertheless they were brothers, they exercised an enormous intellectual influence on
each other, they were raised in the same atmosphere of a highly cultured Russian home, and the very same Vasily Andreevich Zhukovsky had been their
friend since childhood, living at Dolbino for months on end and exerting a
poetic and moral influence on the brothers. And their natural surroundings
were the same, and the people around them were the same—the Russian people. . . .

Ivan Vasilievich Kireevsky is considered the founder of Slavophilism as a
manifestation of nineteenth-century public thought. He was evidently such a
good, positive, and enlightened person that not only his friends—Pushkin,
Zhukovsky, Gogol, Pogodin, Aksakov, Shevyryov, and Turgenev—spoke of
him in exalted terms, but even his enemies, who at that time called themselves
"Westernizers." Even Pisarev—who, as is well known, had attempted in several articles to severely criticize all of Pushkin's lyric poetry and (separately)
Eugene Onegin—even he, in the article "The Russian Don Quixote,"[24] from the
most extreme position of his so-called realistic criticism, wrote about
Kireevsky respectfully and as one deserving of consideration.

It is worth mentioning that the term "Slavophilism" has gradually taken on
a completely distorted and incorrect meaning, almost becoming a pejorative
word. Popular opinion holds that the Slavophiles revered only their own culture and viewed all other cultures and nations with contempt and scorn, if not
with outright hatred. What an unfortunate delusion, what a perversion of the
truth! Of course, in every social, philosophical, or political movement there
invariably arise extreme tendencies (typically the Right and the Left), and in
these extremist positions the original thinking may be distorted and debased.

Concerning Kireevsky, any form of narrow-minded or vulgar view of Western culture, and likewise of the future of Russian culture, were foreign and repulsive to him.

Kireevsky's fundamental position in a somewhat simplified retelling relates to the notion that Western culture is rational and, up to the nineteenth century, had exhausted itself and could no longer give anything new or fertile to humanity. The new word would belong only to the still developing Russian culture, *if it could at that point master everything that the West had already contributed to the world*. The new word would not be rational, but spiritual, and it would come from Russia:

> Considering the current state of thought in the various European countries, Kireevsky notes that the beginnings of enlightenment that have affected Western history have at the present time turned out to be unsatisfactory with respect to new, recently conceived topics of intellectual inquiry. . . . Contemporary enlightenment in Europe must now give way to a new kind of education, as Graeco-Roman education at one time had to defer to European teaching. The new enlightenment must come from the depths of the Russian nation, and to realize this aim we do not have to adopt the features of Western life in toto, and we also do not have to make an abrupt turn to our own past history. The correct path for Russia's development must relate to the fact that we will infuse Western intellectual life with new meaning. . . . Our final goal is an enlightenment which is incompatible with the one-sided enlightenment of the West and which completely meets the needs of the living, reasoning spirit.[25]

> Our literature's dependence on Western literature not only should not deprive us of hope for its independent development, but indeed should promise for our literature a many-sided character. . . . We have hopes and thoughts for the great calling of our fatherland. In order for our hopes to be realized, it is necessary to take a series of measures aimed at the mastery of Europe's level of intellectual life. Western Europe is noticeably leaving the stage in its role as the leading force of enlightenment, and in the absence of other candidates for this role, it is we who must replace her. . . . For the fulfillment of our national aim—to replace a West that has completed its development—we also must raise ourselves to that same level of development, and to this end first and foremost we need a philosophy. . . . But Western philosophy is not suitable for us, since "foreign thoughts are useful only for the development of one's own. . . ." Our philosophy must develop out of our life. . . . Russia has embarked on a developmental path later than those of the Western governments, which has resulted in her being "rich with the experience of her elders." The Western peoples have closed the circle of their development and have attained a one-sided maturity. . . . Russia is capable of leading Europe out of her moral torpor, if she first can absorb West European enlightenment, so

that by doing this she can later influence Europe.[26]

Kireevsky maintained that in an isolated enlightenment limited solely by national features "there is no life and no common good, for there exists neither a progression of ideas nor that success which is achieved only through the combined efforts of humankind."[27]

In the above discussion one can find as many delusions as one wants, but it's impossible to find in it either ethnic narrow-mindedness, or arrogance, or a narrowness of view, or a disdainful attitude toward the cultures and enlightenment of other peoples.

> As a result of the breadth of his Slavophile views, Kireevsky came to occupy in Moscow society a kind of middle ground between two currents. What linked him with the Westernizers was his respect for the best features of Western cultural life, and with the Slavophiles—his profound respect for his native Russian culture. . . . The breadth of Kireevsky's moral sympathies attracted to him people with various convictions . . . they all valued his intellect; moreover, he consistently received only praise for his character and his high moral standards. In 1832 Pushkin wrote, concerning the suppression of *Evropeets*: "The journal *Evropeets* has been closed down. . . Kireevsky, kind and modest Kireevsky, has been portrayed to the government as a madcap and a Jacobin."
>
> "I don't know the Kireevskys," wrote Belinsky, "but judging from the accounts of Granovsky and Herzen, they are fanatics, especially Ivan, but noble-hearted and honest people."
>
> In one of his letters Granovsky wrote, "With all my heart I respect the Kireevskys, despite the complete divergence of our convictions. In the Kireevskys there is such holiness, directness, and faith, the likes of which I have not yet seen in anyone."[28]

But of course we can find the most forceful, illuminating, and fair-minded words about the Slavophiles in Herzen's *Kolokol*, of January 15, 1861:

> The Kireevskys, Khomyakov, and Aksakov accomplished their mission; regardless of whether they lived for a long or short time, in closing their eyes they could quite consciously say to themselves that they had done what they had set out to do. . . . *With them begins a split in Russian thought.* And when we make this statement, it seems certain that no one could suspect us of partiality.
>
> Yes, we were their opponents, but very strange ones. We had the *same* love, but not the *identical* one.
>
> Imprinted in them and in us from our early years was the same powerful, instinctive, physiological, and passionate feeling, which they interpreted as memory,

and we—as prophecy: a feeling of boundless love for the Russian people that embraced everything in them, their way of life, and their cast of mind.

And like Janus or the two-headed eagle, we looked in two different directions, when *the heart beating in us was really the same.*

They gave all their love and tenderness to their oppressed mother. In those of us raised outside our mother's house this emotional tie has grown weak. We were raised in the arms of a French governess and learned too late that she was not our mother; our mother was a worn-out peasant woman, but even so we guessed the truth ourselves from our resemblance to her, and also because her songs were dearer to us than vaudevilles. We loved her fiercely, but her existence was too narrow for us. In her little room we felt stifled by all the darkened images of the icons in their silver framework and all the priests with their parish clergy who frightened the unfortunate woman already oppressed by soldiers and petty officials. Even her eternal lament about her lost happiness rent our heart, for we knew that she didn't have any cherished memories. We knew something else, too—that her happiness lay ahead, that in the depths of her being there beats an embryo that is our younger brother, to whom we will defer without any benefit for ourselves. . . . Such was the nature of our family discord some fifteen years ago. Much water has passed under the bridge since then, and we have met the *mountain spirit* that stopped our flight, but instead of encountering a world of powers, they [the Slavophiles—V.N.] came up against living Russian questions. It is strange for us to consider this, for we don't have a patent on our understanding. Time, history, and experience have brought us closer to each other, not in order for them to convince us to join them or for us to entreat them to join us, but because both they and we are closer now to more truthful views than we were when we mercilessly tormented each other in journal articles, although even at that time I don't remember that we doubted their burning love for Russia, or that they doubted ours.

On the basis of this belief in each other and because of our common love, we, too, have the right to bow to their coffins and throw our handful of earth on the deceased, with the sacred wish that on their graves and ours the young Rus will blossom sturdily and profusely.

I have already mentioned that the Kireevskys' estate, Dolbino, was situated not far from Optina Monastery and that (not only because of this but partly for this reason, too) both brothers were closely connected with this monastery. While Peter Vasilievich, in collecting folk songs, housed his materials in the Optina library, Ivan Vasilievich occupied himself with his publications.

As we know, Paisius Velichkovsky translated a great deal himself with the help of his students and followers, but during his lifetime he was virtually unable to publish anything. All of these translations existed in manuscript form and gradually wound up in the Optina library. Elder Macarius got the idea to organize a publishing house at the monastery in order to publish these manuscripts. He sought the advice of Ivan Vasilievich and Natalya Petrovna Kireevsky and received their enthusiastic support. Ivan Vasilievich even had a little house built at Dolbino in which Elder Macarius and his helpers could prepare the manuscripts for publication, and also work on new translations of ancient books. Through Professor Shevyryov, Kireevsky obtained permission to establish a publishing house at Optina, and the enterprise was launched.

Optina Monastery published all the manuscripts of Paisius Velichkovsky, as well as new translations. Ivan Vasilievich Kireevsky participated most actively in the publication of the books: he did translations and reference work, obtained books that were needed, corrected the proofs, and, finally, gave the publishing house substantial material help.

Naturally, both brothers (and they died only four months apart) were buried at Optina Monastery, near the east (altar) wall of the Cathedral of the Entrance of the Holy Virgin into the Temple (the main cathedral of Optina).

The more significant historical or literary figures are, the more difficult it is to write about them now. Their every step, gesture, and biographical fact have long since been identified, described, and published. The task before today's essayist lies only in reminding the reader, in refreshing his or her memory about specific words, gestures, and facts, since they pertain to the subject that interests us.

It is well known that Nikolai Vasilievich Gogol was closely linked with Optina Monastery, that he loved this place, visited it on occasion, corresponded with Elder Macarius, and also wrote letters to other residents of the monastery.

His first visit took place in 1850. Gogol feared the cold of a northern winter and either didn't want to travel to Rome anymore or was unable to; hence he decided to spend the winter in Odessa. Incidentally, it is possible that this was only a pretext for travel, since in order to winter in Odessa one could leave Moscow even later than the thirteenth of June. However slowly people traveled from one place to another in Rus in those days, it still didn't require four or five months to reach Odessa.

Gogol loved to travel and felt better on the road than he did staying in one place. He knew how to live, think, and work on the road, all the while gathering impressions and resting his spirit. He even had a plan to travel to all the monasteries of Russia, keeping in mind that they were situated in the most beautiful and typical locations. He wanted to store in his soul a virtual collection of the most beautiful landscapes of Russia. Moreover, on the basis of these impressions he intended to write a geographical work about Russia in such a way "that one could apprehend the ties between the individual and that soil on which he was born." Gogol talked animatedly about all of this at a dinner at A. O. Smirnova's in the presence of Alexei Konstantinovich Tolstoi.

This dinner took place on June 16 in Kaluga, which meant that the road from Moscow to Kaluga took the travelers (Gogol and Maximóvich) three days to traverse. On the first day they reached Podolsk, the second night they spent in Maloyaroslavets, while the third was spent in Kaluga.

Maximóvich states (according to the records of Kulish, entitled *Notes from the Life of Gogol*, 2:235):[29] "While on the road Gogol liked to stop at monasteries and pray to God in them. He especially liked Optina Monastery on the Zhizdra River near Kaluga."

When he was about two versts from Optina Monastery, Gogol stepped out of his carriage in order to approach the monastery on foot. As we know, it was June (the beginning of July, according to the New Style calendar), the time of year when wildflowers are at the height of their summer blossoming.

Travelers walked along the well-known broom path or, more precisely, along a road lined with broom plants (white willows) that led across the floodplain of the Zhizdra. In light of the abundant bright flowers, blue sky, mild summer warmth, and white buildings of the monastery capped with gold against the background of the dark forest, one can imagine the feeling of earthly well-being that the sensitive soul of the great artist was experiencing. And just then Gogol saw a girl walking toward him with a bowl of wild strawberries. Gogol wanted to buy them, but the girl gave him the berries for free, saying, "It's not right to take money from travelers." He found this gesture by a peasant girl endearing and moving. In a letter to Count A. P. Tolstoi written about twenty days later, he describes it: "While on the road I stopped at Optina Monastery and took the memory of her away with me forever. Grace is visibly present in that place.[30] One feels this even in the external rituals themselves, although I can't explain to myself why. . . . While one is still several versts away from the cloister one can already smell its perfume: everyone becomes more affable, bows are lower, and sympathy toward others increases" (*Letters*, 4:332).

In the monastery Gogol lived at the skete in a small house that miraculously has remained intact to this day. People say that at the time the entire skete looked like nothing but a flower garden enclosed by a fence, and incidentally, these were among the rarest and most skillfully grown flowers.[31] If one adds to this the quiet and the chime of morning and evening church bells, it is possible to imagine the atmosphere in which Gogol lived. He walked a great deal in the surrounding environs, gathering medicinal plants, but he also read extensively. It is known in particular that at Optina he read a book by Sirin[32] (probably in manuscript form), and that this book made a profound impression on him. That is, it is not even the point that it impressed him, but rather that it prompted him to reconceptualize one of his fundamental opinions about morality and life. This opinion also remains one of the principal, most characteristic contradictions of Christianity as a teaching. The contradiction is as follows.

On the one hand, Christianity (like most religions) states: "Everything comes from God." "Such is God's will." "Thy will be done." "That was the Lord's will." "God has given, God has taken away." "Marriages are made in heaven." "Not even a hair will fall without God's will." And so on. This means that everything in a person's life has been preordained and nothing depends on that person's own actions. Everything comes from God.

On the other hand, the very same Christianity (as distinguished from other religions) proposes and even insistently cultivates the goodwill of an individual and states that a person must fight against his or her own impulses, against sin, and against darkness and filth, in order to attain purity of soul, salvation, and the perfection of the self. It turns out that God is not really an all-powerful and all-controlling supreme phenomenon but rather only an indicator, like a compass or a lighthouse, to show people in which direction they should sail, what they should accept from an enormous and complex world, and what they should reject.

We encounter the pathos of predestination in one part of Gogol's *Dead Souls*,[33] specifically, in the eleventh chapter of the first part: "There are passions whose choice is not determined by the individual. They were already a part of him from the moment of his birth and entrance into the world, and he is not provided with the strength of will to turn away from them. They play themselves out according to a supreme plan. . . . They are called forth for blessings unknown to the individual."

And so you see, after reading Sirin's book at Optina Monastery, Gogol wrote in pencil on the pages of his *Dead Souls* (the first edition), just opposite the above citation:

> I wrote these words in a state of spiritual delusion, and it's all nonsense; innate passions are evil, and a person should concentrate every effort of reason and willpower to eradicate them. Only the smoky haughtiness of human pride could instill in me the thought of the supreme significance of innate passions. Now that I have grown wise, I deeply regret the "foul words" written here. When I was publishing this chapter I felt that I was getting mixed up, for the question of the meaning of innate passions occupied me a great deal and for a long time; it hindered the continuation of *Dead Souls*. I regret that I became acquainted with Isaac

Sirin's book so late; he possessed great insight into the soul and was a perspicacious monk. One encounters a healthy psychology and a correct, rather than a false, conception of the soul only in ascetics and hermits. What young people who are themselves confused by the cleverly interwoven German dialectic have to say about the soul is nothing more than an illusory deception. An understanding of the nature of the soul is not given to the person sitting up to his ears in the mire of everyday life.[34]

The volume of *Dead Souls* containing these marginalia in pencil first belonged to Count A. P. Tolstoi, but, at the very least, several people saw it when it was in the Optina library.

Gogol was always in need of moral and material support. We recall that, while living in Rome, he conceived the fantastic project of having Shchepkin and Konstantin Aksakov come from Moscow to fetch him in order to free him from the petty traveling details related to his return to Russia. "It's distressing and almost impossible for me to occupy myself right now with the details and bother of the road . . . I need to be cared for and pampered," Gogol wrote with a steady pen, as he similarly would write ten years later, hardly having left Optina, to Optina Hieromonk Philaret: "Pray for me, Father Philaret. Ask your worthy abbot, the entire brotherhood, and everyone who prays zealously there [to pray for me—V.N.]. My journey is difficult, and my situation is of the type that without the minutely, hourly, and manifest help of God my pen cannot move. . . . I tell you that every minute my thoughts must be on a higher plane than that of everyday troubles, and at all points of my wanderings they must be at Optina Monastery."

We don't know how it was in his thoughts, but in reality Gogol visited Optina after slightly more than a year, in the fall of 1851, when he had set out for Russia Minor to attend his sister's wedding.

This time his relations with Macarius became almost tragicomic, as we might call them if the matter did not nevertheless have to do with a great Russian writer. Concerning the previous relations between Gogol and Macarius, D. P. Bogdanov writes in his article "Optina Monastery and Pilgrimages Made to It by Russian Writers"[35] (*Istoricheskii vestnik*, October 1910, p. 33):

> Macarius, the elder who had impressed Gogol to the depths of his soul, was a monk of lofty spiritual life. The entire brotherhood of the monastery accepted his advice and directions; for them he represented a tireless instructor on the road to Christian perfection. Gogol's soul was attracted most of all to the lofty, selfless mind of Elder Macarius. . . . According to the recollections of contemporaries, the relations between Gogol and the elder were most congenial. Gogol

asked the monk for his help in resolving all the questions and doubts troubling his soul, while the monk in turn listened to his concerns in a state of friendly readiness, offering advice and instruction.

It was between this particular elder, Macarius, and Gogol that an amusing conflict took place.

The French have a saying that translates something like this: "Don't ask of God what can be done by ordinary jurisprudence." Indeed, obviously one should not appeal to higher quarters for the resolution of questions that one can easily resolve by oneself. Well, all right, if Gogol had been pondering such questions as "Should I go on living in this world?" "Should I burn *Dead Souls* or not?" or, in general, "To be or not to be?" But his only doubt was: "Should I go to my sister's wedding, taking into account that I have already left Moscow and traveled over two hundred versts?"

I think that very often people seek advice, not when they don't know what to do, but when they have already reached a decision in their souls and seek external reinforcement for this decision from an indisputable authority. Even in our everyday lives, when we ask our friends for advice, we secretly expect them to recommend what we ourselves want to do most of all; by their counsel they support us in our own, not yet final, but desirable decision. We can really develop cool feelings toward our friends if now and then they give us advice that sharply contradicts our secret wishes. And the opposite holds true: how appealing to us is the person who, as if having guessed our wish, supports it with his or her words. Yet many times we might have valued even more the harsh advice of a friend later, after several years had passed, rather than that of a friend who had indulged our feelings at the time.

It must be that Gogol didn't want to drag himself across the expanses of Russia that autumn to his sister's wedding, or that he really didn't feel well; but then he could have returned home. But no, he began to put Elder Macarius to the test as if the latter were an oracle. Macarius probably judged for himself simply that the writer should attend his sister's wedding and that, after all, he had already covered part of the distance, and hence he advised Gogol to go.

This advice didn't satisfy Gogol. The thought of not going lived secretly in his soul, and yet he didn't receive confirmation of it from above. The next day he again went to Macarius, saying that he had had a dream counseling against further travel, and that, on the whole, whenever he thought of Moscow he felt calmer inside than when he thought of Vasilievka.[36]

Macarius could have shown resolve and stuck to his opinion. Maybe he would have shown resolve if the matter had not been such a trivial one. Let us recall an episode from history that is to the point, or perhaps not to the point. Dmitry Donskoi decided to mount a campaign against the Tatars (the

Battle of Kulikovo), and the farewell church service was taking place in the Trinity–St. Sergius Lavra. Sergius himself was celebrating the service. He noticed that Prince Dmitry's soul was in turmoil, that the prince was unsure of himself, was wavering in doubt, and that his spirit lacked a firm resolve. Sergius went into the altar, sequestered himself there for a rather lengthy period of time and, upon coming out, lifted up his arms and, in a loud voice, proclaimed: "Dmitry! I have seen your victory over the enemy. . . ."

But that was another matter, and the proportions of it were different. But here we have a trivial matter, and Macarius gave in: "Well . . . if you had a dream . . . and if you feel calmer inside, then return home."

The matter didn't end there. The next day Gogol appeared once again with his doubts, again seeking advice: Wouldn't it be better to go after all? Macarius told the writer to take an icon in his hands and follow the thought that entered his mind at that moment. Moscow came to mind. When Gogol came to question the elder yet a fourth time, the latter in effect drove him away in exasperation and threatened not to receive him in the future. Apparently, he in fact did not receive him, because on the following day Gogol and Macarius sent notes to each other.

> Yet another word, Father Macarius, you, who are close to my heart and soul. After the first decision that I kept in my soul as I was approaching the cloister, my heart felt calm and peaceful. After the second I felt somehow ill at ease and troubled, and my soul was not calm. Why did you say in parting with me, "Is this the last time?" Perhaps all this is due to the fact that my nerves are on edge; in this case I am very fearful that the journey will ruin my health. It frightens me somewhat to find myself ill while on a long trip; especially when the thought will be eating away at me that I left Moscow, where I would not be abandoned if I had a fit of spleen.

On the reverse side of this letter (no one was worried about collecting autographs!) Macarius wrote to Gogol:

> I am very sorry that you find yourself in such a state of indecision and agitation. Of course, if you had foreseen this it would have been better for you not to leave Moscow. I took to heart what you told me yesterday about how peaceful you felt upon thinking of Moscow, and I calmly advised you to return home, but since you again became worried, I was at a loss as to what to tell you. Now you must decide about your journey yourself; if the thought of returning to Moscow makes you feel calm, then you will recognize God's will in this. . . . [37]

Gogol returned to Moscow and, according to the accounts of his contemporaries, in particular Sergei Timofeevich Aksakov, felt sad and upset, especially on October 1, the day set aside for the wedding, which coincided with his mother's birthday. . . .

The next person who comes to mind upon hearing the words "Optina Monastery" is Fyodor Mikhailovich Dostoevsky.

Can one say that fate accidentally linked him with this place? One can put it this way: accidentally, but fate nevertheless.

While the execution of the Petrashevtsy[38] was taking place in St. Petersburg in 1849, on the scaffold next to the twenty-eight-year-old retired engineer-lieutenant Dostoevsky stood, side by side, his friend and compatriot Nikolai Sergeevich Kashkín.

Even today the village of Nizhnie Pryskí can be found three kilometers from Kozelsk. Ordinary peasant huts, white willow trees, a collective farm, a cowshed, an asphalt road through the village.

> Along this road [at that time, of course, not an asphalt one—V.S.] one hundred and fifty years ago an unusual traveler was returning to Moscow from Kiev—the marvelous poet and former governor of Vladimir, I. M. Dolgoruky. Intelligent and observant, he loved and understood architecture, and his descriptions remain of great interest to us. "From Kozelsk to Peremyshl," he wrote, "the road is bestrewn, so to speak, with beautifully constructed manorial lands. One wants to look at the village of Mr. Kashkín more than all the others . A large stone house consisting of three parts, surrounded by a wonderful garden, gazebos, and artificial tents."[39]

This estate had stood almost to this day. To be sure, during the last decades it was not exactly in the condition to be the one "one wants to look at . . . more than all the others," but at least it identified the place. At any rate, during the first half of the 1930s Boris Petrovich Rozanov, according to his own account, would visit this house, which was already half-destroyed. Its windows had been smashed and the doors torn from their hinges, while the wind roamed through the suites of rooms and pieces of glass and plaster lay scattered all over the floor. But, even so, the house—with its walls, floors, and roof—was still intact. Today there remains no trace of the estate.

And, incidentally, Dostoevsky began to visit this same estate after he had served his prison sentence and exile. The entire Kashkín family was highly cultured, enlightened, and active in civic affairs: Sergei Nikolaevich had been a Decembrist. His son Nikolai had been a Petrashevets. Some years later N. D. Kashkín became a music scholar and friend of Tchaikovsky.

While they stood on the scaffold and heard that their death sentence had been commuted to exile in Siberia, Dostoevsky and Kashkín swore not to forget each other. And in fact a correspondence developed between them. In 1878, when Dostoevsky's son Alyosha died, a death with which the father did not know how to come to terms, Kashkín invited the grieving father to Nizhnie Pryskí, and Dostoevsky accepted the invitation. In this way he found himself only a few versts away from Optina Monastery.

He would go to the monastery on foot. He stayed in a small house that had been set aside for him in the skete. The monastery and skete, along with the city of Kozelsk, became principal locations for the action in his greatest novel. The main character and hero of this novel bears the full name of the son he had lost: Alexei Fyodorovich. From this information alone one can conclude that for the author this novel was his most important and most beloved progeny. The monastery, its skete, its visitors, elders, atmosphere, and its significance for that time and for those living in it occupy dozens of pages in the novel, which anyone who wishes may easily reread. It is a pity that Dostoevsky was not fond of describing the setting, the external description of the place of action, in his novels. Psychology and philosophy in their purest form preoccupied the great writer much more than the beauty of nature. And yet, in one place (perhaps the only one in all of his creative work), the artist did not hold back and gave us a description of the landscape, and this landscape is that of Optina. It is worthwhile for us to recall it:

> He didn't even stop on the little porch, but quickly descended the steps. His soul was filled with ecstasy and it yearned for freedom, for space and broad expanses. Above him the vault of heaven unfolded, expansively and boundlessly, full of quiet, shining stars. From the zenith to the horizon the Milky Way, still cloudy, was divided. The crisp and completely still night enveloped the earth. The white towers and golden cupolas of the cathedral sparkled against the sapphire sky. The luxurious fall flowers in their beds near the house were asleep until morning. It seemed as if the earthly silence had merged with the heavenly, and the secret of the earth had touched that of the stars. . . . Alyosha stood, looking, and suddenly, as if he had been shot, threw himself to the ground.
>
> He didn't know why he embraced the earth and couldn't explain to himself why he so irrepressibly wanted to kiss it, all of it, but he did kiss it, crying, sobbing, and watering it with his tears, and in his rapture he vowed to love it, love it for all time. . . . With each passing moment it became evident and almost palpable to him that something firm and unshakable, like the firmament, was entering his soul. It was as if an idea had formed in his mind and had already established itself for the rest of his life, forever and ever. . . . He fell to the earth a weak young man, but stood up a resolute warrior for the rest of his life, and he sensed and realized it suddenly, at the very moment of his ecstasy.

It is impossible to equate authors with their heroes: Pechorin is not Lermontov; Evgeny Onegin or Hermann are not Pushkin; Levin is not Tolstoi. . . . But if Lermontov had not traveled by post chaise from Tiflis, then Pechorin, too, would not have traveled that same road; if Pushkin had not passionately loved to play cards, and if Tolstoi had not loved to cut grass with a scythe, then Hermann, too, would not have perished because of cards, and Levin, too, would not have found himself cutting grass in a field.

It is really not Alyosha Karamazov at all, but the creator of the novel himself, completely worn-out from penal servitude and doubts, lacerated by psychological and moral contradictions, and having lost his beloved son, who finally "fell to the earth a weak young man, but stood up a resolute warrior for the rest of his life. . . . " And this took place in the heart of Russia, in the maelstrom of her contradictions, and in the full flower and brilliance of her beauty—at Optina Monastery of Kozelsk.

It is thought that Lev Tolstoi visited Optina six times, but in fact there may have been more than six visits. Why is this? We take for granted that everything is known about Tolstoi, down to his last step, his every word. . . . This may be so, but there still remains the question of his early childhood.

Tolstoi was orphaned early in his life and was left without a mother. His guardian, who was his father's sister, Alexandra Ilyinichna Osten-Saken, took the place of his mother and, moreover, replaced her not just formally but essentially, in her tenderness and warmth toward him. Without any exaggeration we can consider this spiritually rich, kindhearted woman the second mother of Lev Nikolaevich. And so, to continue, she loved Optina Monastery very much and probably took her nephews there for visits.

The first documented trip to Optina by Tolstoi took place in 1841, when he had before him the task of burying there, near the Cathedral of the Entrance of the Holy Virgin into the Temple, Alexandra Ilyinichna, who was so near and dear to his heart. It is worth noting that on the gravestone of Countess Osten-Saken are carved the verses of the thirteen-year-old Lev Nikolaevich—one might say, his first promulgated literary work. The poem is of very poor quality, and not because its author was just thirteen years old. Even later, at the height of his literary strength, Tolstoi tried to write poems; but they didn't turn out well. And this is not a pity, for it is enough that he composed what he did in prose. Here are his verses on the gravestone:

> For this world you have fallen asleep,
> And you trod an uncertain path.
> Your life is now in the heavens,
> And one envies your rest so sweet.

In the hope of a sweet reunion
And belief in life beyond the grave,
Your nephews have erected this reminder
To honor the dust of the deceased.

And so, when we consider Optina Monastery and Lev Tolstoi, we must keep in mind not only the literary side of this subject—not only, say, that "Father Sergius" was written with its entire monasterial setting taken from the popular prototype of Optina—but also the fact that, starting in his childhood, distinctively personal moments linked Tolstoi with this place. And we must keep in mind, too, that his beloved sister Maria Nikolaevna lived as a nun not far away, at Shamordino.

Lev Tolstoi's trip to Optina in 1877 while he was working on *Anna Karenina* is linked at this time more with a visit to Beryozichi, Obolensky's estate, and we shall save a description of this trip for a later time, specifically for when we visit Beryozichi ourselves. But now let us turn to a most interesting event in the biography of the great writer, to his walking trip of many days or, to be more precise, his pilgrimage from Yasnaya Polyana to Optina Monastery in 1881.

When one thinks about this trip, unrelated thoughts enter one's mind.

Jesus Christ (according to the legend) brought his teaching to us two thousand years ago. Perhaps his message caught on so fast, and spread to so many places and for so many years, because it was (according to the legend, of course) reinforced by action and fertilized by a sacrifice—by torments on the cross, and subsequently by death.

Tolstoi, too, had a message. If Christ, according to the teaching that we call Christianity, revised the work of the ancient prophets, Tolstoi attempted to revise Christianity—more precisely, not Christianity as such, but the church that had been founded on this teaching and, according to Tolstoi's convictions, had distorted this teaching and strayed far away from it. Even so, this was not simply Tolstoi's attempt to return to Christ in His pure form, but rather his description of these teachings in his own accent. Churchmen were most distressed at the fact that Tolstoi in effect had read the Gospel with pencil in hand, crossing out, underlining, providing commentary and even additions. The sacred, canonical, unalterable text! The classic work of Christianity. . . . One may not correct or add to a classic. . . . This is indeed a fundamental principle!

However, Lev Tolstoi underlined, crossed out, made additions and corrections, and ultimately created his "Gospel," which was published.

Tolstoi also had followers, students, Tolstoians. Tolstoian colonies and brotherhoods were organized. Even today one can still encounter Tolstoians in various countries. Portraits of our broad-bearded Lev Nikolaevich hang in their houses almost like icons.

But his closest students and followers, such as Chertkov, understood that for a firm foundation and rapid development both sacrifice and heroic deeds are necessary. They imagined such a heroic deed to be Tolstoi's departure from Yasnaya Polyana, from his family and his everyday life, in short—a departure. And, of course, if Tolstoi had left at the right time to travel around Rus with a knapsack and in bast sandals, and had wandered about for two or three decades, suffering, enduring privation, preaching, and by personal example reinforcing his sermon, then no one knows what resonance this act of his would have had, what effect his wandering would have exerted on people's minds and souls.

Chertkov persisted, while Tolstoi could not reach a decision, although the consciousness of the necessity of a sacrifice always lived in him, and its unrealization evoked in him feelings of irritability and dissatisfaction with himself. A single decisive act, remaining unrealized, was replaced by many minor acts and half-measures, such as the refusal to accept honoraria, passing judgment on those closest to him, and an eternal dissatisfaction with his way of life. But many minor acts cannot replace a single significant one, in the same way that one cannot jump over an abyss in two or three stages. In place of a single spurt, what developed was a decades-long process of slipping and sliding; what should have been an explosion proceeded as a gradual sputtering. . . .

The craving to suffer for his idea slipped out, too, in a conversation with Konstantin Nikolaevich Leontiev at the very same Optina. Konstantin Nikolaevich was living there, occupying a small house on this side of the monastery's wall. Here is their conversation, from the notes of Leontiev himself: "He is incorrigible. He was polite, but argued for two hours. During the conversation I said, 'It's a shame, Lev Nikolaevich, that there's very little fanaticism in me. For I ought to write to St. Petersburg, where I have connections who would see to it that you were exiled to Tomsk, that neither the countess nor your daughters would be allowed to visit you, and that very little money would be sent to you, for you are positively harmful.'"

Konstantin Nikolaevich probably thought that Tolstoi, as the saying goes, was living too comfortably and complaining, and that it is easy to complain when one has a fortune, an estate, and a secure family life, and when one can live in a refined, aristocratic manner. But, you know, if he were exiled to Tomsk. . . . But this was precisely what corresponded to Tolstoi's secret dream of a sacrifice, of suffering for the sake of an idea. He exclaimed: "My dear Konstantin Nikolaevich! For God's sake, write to them and ask them to send me away. This is my dream. I do everything possible to compromise myself in the eyes of the government, and it's all in vain. I implore you to write to them."

The constant and conscious compromising of himself in the eyes of the government—this, too, was a half-measure, a minor act to compensate for the unrealization of the main, most important, one.

And so, there was the dream of a sacrifice, and halfhearted attempts even took place. There exists a photograph of Tolstoi in a peasant's cloth coat with a knapsack on his back and a stick in his hand. Why should the count dress up in the clothing of a wanderer and a vagabond? Pushkin and Gogol, Turgenev and Nekrasov, Bunin and Kuprin traveled great distances around Russia, but they moved from one place to another using the usual means of transportation for that time, and if they traveled on foot, they wore their usual clothing.

Tolstoi's pilgrimage to Optina in 1881 was, in my opinion, also a halfhearted measure. He wanted to see what it would be like in reality. The factual side of this trip is well known, even from the notes of Tolstoi's servant Sergei Arbuzov, who traveled with him. One of the first scenes is as follows:

> They brought the bast sandals at nine o'clock, and I took them to Lev Nikolaevich and asked whether he would put them on immediately or whether he would walk to the city of Krapivny in boots. Lev Nikolaevich decided to put on the bast sandals right away, and gave instructions for the peasant man who had made two pairs of them to be paid thirty kopecks. The countess entered the study carrying a sack made of simple canvas; with my participation, the count put on the bast sandals according to all the rules of the peasant's art, with leggings, and he tied them onto his legs with twine. . . . Next they helped us arrange on our backs the sacks containing our things. . . . For incidental expenses Lev Nikolaevich gave me twenty rubles, but I don't know how much money he decided to take with him. . . .

The servant also didn't know that, although bast sandals were well and good, the count's visiting card was also taken along, to be on the safe side.

It turned out that even the servant had the idea that, at least to Krapivny (and, it follows, in general) they could wear their boots. The boots were taken along, but they remained in the canvas sack. The count wanted without fail to walk about Russia in bast sandals. More comfortable? More hygienic? Hardly. Moreover, unworn bast sandals irritate the feet much more than boots that have been broken in. And, in fact, through the leggings the twine, it turned out, cut into their legs painfully.

The servant reminisces in his notes about the villages they walked through, where they spent the night, and what they ate and drank (mostly milk and eggs, accompanied by the samovar, a necessity in those years), but what could a servant know of the count's keen suffering when, as if from the outside, he looked with the eyes of an ordinary passerby at the estates to the right and to the left of the road. When in those very estates lived all his relatives and good friends. . . .

On the one hand:

> . . . at the edge of the village a woman was walking . . . she was carrying two pails of water on a yoke. After saying hello to us, she asked where God was taking us and whether she should pray for us; the count answered affirmatively. . . . "So you plan to remain in the monastery forever?" "I don't know, maybe," the count answered.

On the one hand:

> "Granny, let us spend the night here," the count asked her. "Dear sir, I'd be glad to let in travelers, but there's no place to lie down: in the gallery over your heads the flies wouldn't let you sleep, and it's hot there, too, and we don't have any beds."
>
> "We don't need beds," protested the count. "Just bring a bundle of hay to the porch for us, and we'll go to sleep there; would you happen to have a samovar, some milk and eggs?"
>
> "Yes, I do, dear sir. . . . "
>
> The old woman treated us cordially and with humility, and it was evident that she liked to take in and shelter travelers.

On the one hand:

> "If you please [asked the village elder—V.S.], what are you inquiring about here? Do you have a document? Because I know many old men—hypocrites of this sort. Come on, show your document."

All of this, on the one hand, was well and good. But, on the other hand:

> As we approached the village of Spasskoe, Lev Nikolaevich pointed out the manor house to me, which had belonged to Sofya Andreevna's [female—V.N.] cousins.

But on the other hand:

> "Here's my document. . . . "
>
> Vasily read, "Count Lev Nikolaevich Tolstoi" and quietly told his father. As people hide under enclosed spaces from thunder and lightning, so the words

"Count Tolstoi" drove away absolutely everyone in just a few minutes, as if from a rainstorm: the village elder and his son, the peasant artel, the workers, and the old woman-supplicant. Only the count and I were left on the porch.

And now let's imagine that his departure was genuine, his sacrifice and heroic deed were genuine, and that all the ties were broken, all paths were cut off, and there was no document attesting to his nobility. How would the scene with the village elder have ended? After all, people were also exiled to Siberia for vagrancy, and moreover they were whipped first too. For example, before he was sent to Siberia, Fyodor Kuzmich,[40] a Siberian elder (regardless of what mysteries he kept to himself), was said to have been whipped thoroughly. It would be a tragedy if this had really been Alexander I, but what can you do? Once he had begun. . . . [41]

But no, the halfhearted attempt and trying on for size of a peasant's way of life were cautious and almost frivolous: here he was in bast sandals, but with a count's document in his pocket. Even his servant didn't notice when Lev Niko-laevich managed to give his visiting card to a monk at Optina Monastery. The course of events was as follows: they weren't allowed to have dinner in the refectory but instead were seated with the beggars. The count tolerated this. Then, " . . . the monk, seeing that we were wearing bast sandals, didn't give us a room, but sent us to the common lodging house, where it was very dirty, and there were insects, too." At this point the servant gave the monk a ruble and was able to coax out of him a small, third-class room, where they had to sleep with a shoemaker from Bolkhovsk District.

> The shoemaker soon fell asleep and began to snore fiercely, so that the count started from fright and said to me, "Sergei, wake up that man and ask him not to snore."
>
> The awakened shoemaker answered reasonably, "What, you're ordering me not to sleep the entire night on account of your old man?"
>
> Soon the monks found out somehow that within the walls of their cloister was Count Lev Nikolaevich Tolstoi. Two monks were sent by the archimandrite and Father Ambrose in order to fetch the count's things and invite him to the first-class hostel, where everything was upholstered in velvet. For a long time the count refused to go there, but even so, finally he decided to go.

Well, subsequently everything proceeded the way it was supposed to: an audience with Elder Ambrose, walks in the early morning dew, bathing in the Zhizdra River twice a day, drinking water from the sulphurous Pafnutius Spring, conversations with the peasants during his walks, but now without disguise and without fanfare.

193

In a letter to Turgenev, Tolstoi exalts: "My pilgrimage has come off splendidly."

What tormented Tolstoi his entire life possibly spurred him on to take that final step now called "Tolstoi's departure." Ultimately, the step was taken after a fateful delay of thirty years. Whether by the convergence of circumstances or by the will of fate, Tolstoi departed for this very same Optina Monastery.

Debates took place (and even today one encounters them, if only as a hypothesis) concerning whether or not Tolstoi wanted to be reconciled with the church and spend the rest of his days at Optina. But here emerge two separate problems that shouldn't be confused. Tolstoi himself resolved both of them in a single sentence well known to everyone: "I would carry out the most difficult penance if only I wouldn't be forced to attend church." His first words are worth considering, too, when the host of the monastery led the runaway into a spacious room containing two beds and a wide sofa. "How pleasant it is here!" exclaimed the man who was exhausted from the journey (and from his entire preceding life), who had shivered from cold in a cheap train, the eighty-two-year-old man who had been bounced around by coach drivers on bad autumn roads. It was not the church he sought, but rather peace, if only physical comfort at first. But he needed something else, for mere physical comfort could be found at Yasnaya Polyana too—in his own study, on his own sofa. And even though it was pleasant here, still (as Makovetsky attests), "the night was interrupted first by cats that would run down the hallway . . . then a woman whose brother had died today came out into the hall and began to wail. . . . " No, if one speaks only of peace, it was much more peaceful at home. But still, there was something that would not let him rest, that moved his spirit, pulling and pushing him, until at last it forced him to run away. And no matter how many times people said that during these days the atmosphere in the house at Yasnaya Polyana had become more tense, let's believe the eyewitness account of Tolstoi's own son, Ilya Lvovich: *"the dream that he had cherished for so long* [my emphasis—V.S.] about leaving Yasnaya Polyana became the only way out for him."[42]

At that time Elder Ambrose was no longer among the living: for nineteen years already his remains lay buried near the Cathedral of the Entrance of the Holy Virgin into the Temple. In all probability he would have been the first to take the initiative. Elder Joseph, who had replaced Ambrose, waited for Tolstoi to come, while Tolstoi was waiting to be invited. And in the meantime, their desire to meet each other was obvious. While still on the train, Lev Nikolaevich queried which elders were presently at Optina and asked the carriage driver the same questions as they were riding from Kozelsk; he told Makovetsky that he was going to visit the elders.

But the next morning he began to have doubts. Chertkov's secretary,

Sergeenko, came to see him, which for all intents and purposes was the same as Chertkov himself coming to visit. In front of Sergeenko, Tolstoi said that he would not go to visit the elders. But Sergeenko soon left, and for some time afterward one could say that Lev Nikolaevich "circled" the skete and Elder Joseph. On one of his walks he went directly to the skete (to its southwest corner), walked along its southern wall, and disappeared into the forest. After eleven o'clock in the morning he took another walk, and once again went to the skete. He reached the Holy Gates, went back in the direction from which he had come, considered turning to the right, returned again to the Holy Gates, and then went behind the tower to go once more—toward the skete.

Here he encountered a monk holding a broom (Pakhomy) and, entering into conversation with him, stated in plain terms that he was Lev Nikolaevich Tolstoi, that he was going to see Father Joseph the Elder but was afraid to disturb him, since he had heard that the elder was ill. Pakhomy answered that the elder was not ill, only frail, and that of course he would receive Lev Nikolaevich. Tolstoi set off to see the elder, but upon reaching the Holy Gates he again turned suddenly and went into the forest. When he returned to the hostel after wavering so many times, he said to Makovetsky: "I won't go to visit the elders first. If they were to invite me themselves, I would go."

Pakhomy (on whom the meeting with Tolstoi had made an enormous impression) was later rebuked for not taking Tolstoi at once to Joseph. Pakhomy said in response that he just didn't dare.

Dushan Petrovich Makovetsky, for many years personal physician to Tolstoi as well as his secretary and in general a very close friend of his, in my opinion dots all the *i*'s in saying: "Lev Nikolaevich evidently wanted to talk with the elders very badly. I interpret his second walk (L.N. never took two walks in the morning) as his intention to visit them. . . . L.N. wanted to see the hermits-elders not as priests, but as hermits; he wanted to talk with them about God, about the soul, about asceticism; he wanted to see how they lived and find out under what conditions one could live at the monastery."[43]

Incidentally, at this point one can only make conjectures. The meeting did not take place, and Tolstoi left Optina to go to Shamordino, where Maria Nikolaevna lived. He was in a hurry. Locked into his own subjective impressions of himself and of the correlation between himself and reality (as often happens in people's minds), he couldn't evaluate the situation realistically. In a burst of naive conspiracy, he signed a letter to Alexandra Lvovna "T. Nikolaev," planning to run away and hide when the entire world had already been keeping track of every meter of his movements around Russia.

But this time he didn't have any visiting cards. The innkeeper Mikhail asked for a card, saying that it meant something to him that such a personage had stayed at his inn—but there was no card. The only remaining alternative was to sign the guest book. And these were virtually the last written words of Lev Tolstoi: "Lev Tolstoi. Thank you for your hospitality."

II

It is well known that ideas are floating in the air. When the time comes for something to be invented, the invention occurs almost simultaneously in different places, by several people working independently of one another, but depending on a common, fertile environment or atmosphere, on the fact that "the time has come."

Such was the case with the steam engine, the electric light bulb, the radio, the atomic bomb, and the appearance of many hypotheses and theories.

There are examples of lesser scope in our sphere as well: similar coincidences have occurred in journalism, in publicistic writing, in art in general, and in the area of the civil and patriotic movements of the soul.

For some time now Optina Monastery had become for me a subject for publicistic investigation: I had heard of the monastery, of course, and in the past had even known something about it in general terms, but somehow it was all general and extraneous. And suddenly, the stars, if you will (in jest), aligned themselves with each other appropriately, and the logic of the writing table (in earnest) attracted me; but increasingly thoughts began to emerge about this historical, literary, cultural, architectural, and, ultimately, national treasure.

And suddenly—a convergence of ideas. Suddenly I see that the art historian Vladimir Desyatnikov has published an article, "Optina pustyn" (Optina Monastery), even if only in the Zagorsk local newspaper. And then Dmitry Zhukov, a veteran fighter for Russian cultural treasures, suddenly publishes an outstanding article in *Literaturnaya Rossiya* entitled "Kto vosstanovit pamyatnik?" (Who will restore this monument?). A letter arrives at about the same time, and not just a letter, but a packet containing a manuscript of seventeen typed pages. The writer is someone unknown to me, A. E. Sazykin of Smolensk Province. In the letter the author introduces himself: "I am a military man and have served in the army for twenty years. In four years I plan to enter the reserves and shall have plenty of free time to travel, explore, and read." In this letter Officer Sazykin (it's a pity that his full name is not written out) writes:

> Optina of Kozelsk—the former monastery—is an extraordinarily interesting place that is popular in literature, but there is almost nothing to see there, and not everyone can imagine what it used to be like. In recent times no one ever preserved anything there, including the buildings renovated by the SPTU,[44] and these decayed in time, while what was of little value people destroyed, out of

196

mischief or greed (a piece of iron was stolen from the cupola of the cathedral . . .). Very little remains: the framework of the churches, two or three small towers, part of the fence . . . we search for signs of extraterrestrial civilizations, but wipe away the traces of our own. . . .

This picture corresponds with Dmitry Zhukov's description of Optina Monastery (*Literaturnaya Rossiya*, Jan. 20, 1978):

> Upon opening my eyes I saw neither the white towers nor the gold domes of the Cathedral of the Entrance of the Holy Virgin into the Temple. The war, time, and neglect had performed their fearful, destructive work. Dirty ruins that are not picturesque at all surround several intact buildings too small for the rural professional-technical school housed in them. Piles of broken bricks alternate with the rusting parts of combines abandoned as being of no use. . . . Jumping over debris, I make my way to the graves of the Kireevsky brothers enclosed by a pitiful fence. . . .

And so, events began to coincide with each other. Yet another letter: "Excuse me for the intrusion . . . I was told that you were planning [I was still only thinking about it, but someone had already uttered the words!—V.S.] to visit Optina Monastery. . . . To aid you in your undertaking I recommend that you contact the director of the Kozelsk Regional Historical Museum, Honored Cultural Worker of the R.S.F.S.R. and journalist, Vasily Nikolaevich Sorokin, who resides in Kozelsk. . . ."

Several days later something else happened. A letter arrived from none other than Vasily Nikolaevich Sorokin himself: "I heard that you were planning to come to Optina Monastery. I applaud your decision and ask you to let me know when you plan to realize your intention. It would be good if you were to do so this summer. We'll take care of your lodging. . . . "

At six o'clock that evening I try to order a call to the city of Kozelsk. Thus far the city had been just a name to me, aside from the few historical facts I knew about it. But I know that it's not very far away, beyond Kaluga, beyond Peremyshl (and Peremyshl is also just a name to me), probably about . . . well, anyway, less than three hundred kilometers. And in that city there lives a certain (also just a name to me so far) Vasily Nikolaevich Sorokin, director of the museum, Honored Cultural Worker of the R.S.F.S.R., about whom Dmitry Zhukov had written so warmheartedly in that same article:

> And they come, they come without stopping. From all the corners of the world[45] people come to this place that is not included in booklets for tourists. And Sorokin, an elderly, gray-haired man, gets into the bus and, burning with shame

about the ruins he would now have to show us, quickly begins to speak about people who are world-renowned. Having sacrificed a career as a journalist and writer for the sake of preserving even a small part of the remains, he speaks elegantly and with enthusiasm . . . diverting our attention away from the ruins of the fence, corner towers without roofs, churches, bell towers, and buildings without roofs and with gaping holes instead of windows. . . . He hurried along the guests, who were bogged down in mud up to their ankles, to the skete, where a relative orderliness had already been established. . . .

And now, within an hour (according to the existing rules of our system of communication) I'll hear this man's voice and we'll decide on the day of my arrival.

The telephone operator briskly answers: "You can reach Kozelsk only after 1:00 A.M."

"But why? Before 1:00 A.M. one can reach New York, Addis Ababa, and Rio de Janeiro. . . . "

"You can reach Kozelsk only after 1:00 A.M."

"But in that time I could drive to Kozelsk and maybe even drive back."

"Have a nice trip!" The telephone operator hung up.

And that is how I learned that Kozelsk lies somewhere outside of the mainstream of civilization, since it really is true that nowadays one doesn't encounter any problem in contacting practically any city in the world by telephone. I suspect that I could have reached London or Stockholm from Moscow faster than Kozelsk.

At five o'clock in the morning I heard the voice of Vasily Nikolaevich. The first thing I did was begin apologizing for such an early phone call, but it turned out that Vasily Nikolaevich was already awake at that hour. We decided that on Friday, July 21, I would arrive in Kozelsk.

I asked Volodya Desyatnikov to come with me, and he readily agreed. He had already been to Optina, had written about it, and was well acquainted with Vasily Nikolaevich. An art historian, he had been studying Russia's distant past (her remains) for a long time, and he knew what could be found where, had traveled extensively, and had seen everything with his own eyes. My goal here lay not in the bare facts (a fashionable term these days) that I would have obtained during the trip, but in the fact that with Desyatnikov there could be no other conversation on the way except about the topic at hand—only about churches, monasteries, icons, and books. This way, right from the moment of our departure, we began to live and breathe the topic for whose sake we were going.

Volodya brought along for the road a little book bound in yellow with a depiction of Optina Monastery on its cover. The book was entitled *The Environs of Kaluga.*

"Is that a guidebook?" I asked.

"Don't you know this? It's Evgeny Viktorovich Nikolaev's book."

I didn't know it.

It was a marvelous little book. Very informative. Written lovingly. And the main thing is that he understood everything.

"What do you mean 'understood?'"

"He died. And he was only thirty-two. By profession he was a scientist, a chemist. But what insights he had into art and beauty! . . . Listen, let me read you something. I haven't read this anywhere else, and not by anyone else. . . ."

Being well acquainted with the book, Desyatnikov turned several pages and read:

> The crosses of Zagorie resound with the music of a spring holiday. This represents in general one of the brightest pages of our fine arts, and to date it has not been sufficiently appraised. Without the lace of the crosses one cannot imagine a single church of the end of the seventeenth century, including Zagorie—they are not merely the completion, but a most important detail of its image. The central cross—a huge object—measures three meters, a size that is deliberately exaggerated in proportion to the cupola, but from below it resembles a piece of jewelry.

"Yes, I'm really hearing about this for the first time. Of course, I always understood that to restore a church and at the same time remove its crosses is stupid, tantamount to knocking the head off a work of sculpture or shaving the hair off a woman's head, depriving her of a hairstyle and forcing her to appear in this way. If one considers a church a work of art, then the replacement of its crosses with pointed and featureless pins results in the destruction of the whole, and, at the very least—ignorance. . . . And, after all, it's true that crosses represent an entire branch of architectural art; your Nikolaev was right. . . ."

"Yes, you see what I mean. He understood everything. Or take this example. What does a ravine mean to a city, to Kaluga, for instance? An absurdity—it spoils the view. Send it to the devil, fill it in. But listen to what Evgeny Viktorovich writes about this ravine: 'In an act of amazing tactfulness they preserved the ravine. Even today it remains one of the most charming places in the city. Green slopes descending to the river, an unevenness to the relief—all of this seems completely natural, but behind it lies a profound understanding of the city's artistic appearance.'

"Or listen to this about Peremyshl: 'In 1777, having become a district city, Peremyshl, along with other cities, received a new plan. It was a good plan, according to which the city was made "regular," but they didn't commit any kind of violence against its relief. The center was kept where it had been, above a precipice, while the city's landscape remained intact.'

"If only our architects understood the aims of city planning the way this chemist did. And the way he writes about country estates, even about what remains of them! Listen to what he says about Gorodnya, the oldest estate of the princes Golitsyn: 'The house in Gorodnya is emphatically simple. One thinks that one can understand it instantly, but as one considers it for a longer time, one is struck by the many rich nuances of its architecture. The artistic placement of the facades enhances the refinement of its conception. . . .'"

At this point I began to laugh.

"Why are you laughing?"

"I was imagining how the people who acquired all these buildings for their economic needs—the directors of the MTSs, of state collective farms, even rest homes, too—considered the architectural nuances and the refinement of the conception."

"Let me read on":

> From the road one can see the house in its interrelationship with nature. The calm, simple house appears poetic from here. Established high above a pond, it reveals its monolithic self slowly: now one sees it in the clearing between trees, now it is hidden by them, while its hazy reflection shimmers in the water. . . . The builders avoided "frontal" details, and hence one can't gaze enough at this sight. One knows perfectly well that all of this was "made" and calculated, but still it seems absolutely natural.

"But what can be found these days at this place, at Gorodnya?"

"One can imagine the conception. But in other respects . . . the frame of the house is intact. The park has been neglected and has lost its form. . . . No, you read Nikolaev. Read how he writes about Polotnyanyi Zavod of the Goncharovs,[46] about Grabtsev, about Avchurin. There is so much historical information, such subtlety of observation, so much love and pain. . . . And what a description of these country estates is offered by Nikolai Nikolaevich Voronin, the marvelous scholar and fighter for the preservation of Russian culture. . . . I quote: 'You see, only in the quiet world of the country estates' gentle beauty (with their parks) could such giants of the Russian genius as Pushkin, Tolstoi, Turgenev, and Goncharov, with their heartfelt love for the beauty of the earth and humanity, be born and nurtured.'"[47]

"But why is it possible, and so easy, to ruin all this, while to repair it is impossible? Or don't we want to? Say, to repair that same Polotnyanyi Zavod of the Goncharovs. The parks, system of ponds, the Great Garden, Lower Garden, the enormous palace. And, in my opinion, it doesn't matter that Pushkin had been there, or that in 1812 the headquarters of the Russian army was located on this estate."

"It does matter."

"No, it doesn't. For us there always has to be some memorial 'hook.' But if we act only out of other artistic considerations—of architecture, parks, land-scaping—then a wild place will be transformed into its own kind of work of art, into a wonderful work of art. Why should they neglect it and leave it in a defaced state? . . . How many monasteries were there in Russia?"

"One thousand and eighty."

"Isn't it well known that monasteries were built in the most beautiful loca-tions and that later on they inscribed themselves into the landscape, comple-menting and beautifying it?"

"Yes. Gogol even wanted to travel to all of . . . "

"That's well known, too. So who would be offended today if in our country there existed one thousand and eighty beautiful places, virtually the same thing as one thousand and eighty works of art? How many country estates like these were there about which Evgeny Viktorovich writes? Take the famous Sofievka in Ukraine."

"I can cite some concrete facts about Sofievka."

After Volodya had rustled through the papers in his art historian's briefcase, an album appeared in his hands as if by magic.

"Here it is. *The Miracle of Umansk*. By A. P. Rogotchenko. Published in 1977 in Kiev. Testimonies of contemporaries, a description of its splendors, exalta-tions. 'Come to Sofievka and marvel at the artistic genius in it. Here nature and art, in combining all of their forces, have produced a magnificent creation. . . . The garden is one of the happiest and most indispensable consequences of the art of humankind.'"

"I see. It's been preserved and we take tourists there."

"Yes, it even says here . . . I quote: 'A multitude of visitors comes here from thousands of kilometers away—from Moscow, Leningrad, Minsk, Vladivos-tok, Kamchatka. . . . Sofievka receives tourists from Poland, Bulgaria, Yugoslavia, Mongolia, Italy, Canada, and the U.S.A. . . . And people have not been disappointed in their expectations: Sofievka presents all those who arrive with such rich impressions.'"[48]

"One can say the same thing as well about other country estates that remain intact: Arkhangelskoe, Kuskovo, Ostankino. . . . Well, perhaps not of such magnitude, to a lesser degree. But if this is real beauty, do the proportions mat-ter? It only matters that it be beautiful."

"That's right. And the beauty is everywhere—in the house, garden, ponds, arbors, and stone stairs. Can you imagine what the earth would look like if all of this were to be restored to its original condition?"

"It would be impossible to restore all of it, we don't have to entertain such big ideas. At least a tenth of it, at least the main part. Sofievka was not an exception. Let's assume that the Pototskys could allow themselves a little bit more than some landowner of Oryol—they certainly were nobody's fools. Not long ago I received a letter from Oryol, sent by a scholar in the field of landscape architecture, Natalya Olegovna Levitskaya. For the purposes of her dissertation she was developing a project to restore Kireevsky's country estate."

"What do you mean 'Kireevsky'? Which Kireevsky?"

"No, Ivan and Peter aren't related to him. This one was named Nikolai Vasilievich, and he was just a nobleman, a landowner. He kept an open house and had a passion for hunting; Tolstoi and Turgenev were fond of visiting him. This Kireevsky, incidentally, was Tolstoi's prototype for the vivid image of the uncle who said 'That's it. March!' in *War and Peace*.

"Natalya Olegovna Levitskaya's work is entitled 'The Park Ensemble in the Town of Shablykino of the First Third of the Nineteenth Century.' All of this is located eighty kilometers from Oryol. The park is the largest in the region, and the remains of its former layout with the old trees and ponds have been preserved in it. Its architectural features have not been preserved. As a curiosity, only one large cast-iron frog remains intact. There used to be a fountain. Archaeologists are able to reconstruct the entire appearance of an extinct animal by means of a single bone that they have found, and similarly one can imagine what the fountain and park looked like because of this cast-iron frog.

"Levitskaya calls the park 'a chef d'oeuvre of landscape art.' A detailed description of the estate exists in a nineteenth-century journal, and also in an album of lithographs with views of the park; finally, there is a project to restore this complex that includes detailed drawings. Why all of this fuss? Why not undertake this project, why not restore it? And, moreover, the same kind of album as the one about Sofievka, and the same words: 'Every year hundreds of thousands of people from Moscow, Leningrad, Poland, Italy, and the U.S.A. come to Shablykino.'

"Isn't it true that the following restored places beautify our earth: Vladimir and Suzdal, Rostov the Great and Pereyaslavl-Zalessky, the Trinity–St. Sergius Lavra and the Alexandrov Sloboda, the Ferapont Monastery and Kolomenskoe, Pavlovsk and Peterhof. . . . Why couldn't we as well organize and restore Solovki and Valaam, Torzhok and Solotcha, the Tikhon and Sarov hermitages, New Jerusalem, dozens of other monasteries, hundreds of country estates, and thousands of village churches and bell towers that, after all, also beautified the earth?"

"New Jerusalem is being restored. A huge effort is under way."

"I know. I know that around Moscow, especially near the streets, little churches have been put in order that for decades looked depressing because of their run-down appearance, but now they sparkle and shine like jewels. I know that many monuments are surrounded by scaffolding, and it appears as if

202

restoration work is proceeding. Oh, those scaffolds! They survive afterwards for ten or fifteen years. . . . But even that's good.

"It's good, it's very good that much is being accomplished already, but it's bad that this is proceeding only in individual cases, as if they were exceptions, in accordance with separate exceptional resolutions. Suzdal has been completely restored, but the Florishchev Hermitage in the same Vladimir region and the Spas-Kupalishche remain in their former condition. Concerning monuments of the past, there doesn't exist an overall favorable and benevolent atmosphere, an overall favorable climate."

"You're not entirely correct. The overall atmosphere has changed and continues to change for the better. Remember that not so long ago there didn't exist a Russian Society for the Preservation of Monuments of the Past, but now it exists. A special law has been passed to protect historical and cultural monuments. An article concerning the protection of monuments has been written into the new constitution.

"Here we are, traveling to the Kaluga region, in which currently fifty monuments are being restored. In 1972 the Kozelsk division of the Kaluga restoration workshop was founded. Since then—that is, since 1972—it has appropriated 164,000 rubles. And in 1978, out of an anticipated 44,000 rubles for Optina Monastery, they received 41,000."

"And for the entire country?"

"Fifty million."

"That's a tidy sum, too, of course. But you understand yourself how little 40,000 rubles can do for Optina Monastery. That's the price of a dacha outside of Moscow. And if you rebuild everything from the ground up, you won't make it. And fifty million for the entire country. . . . "

It turns out that we, each of us, and even the Society for the Preservation of Monuments of the Past, are fighting for each monument separately, fighting for a single little church, and, as it happens, we save it. But is it right that we have to fight for them?

Our car rushed on, and on both sides of the road the earth turned slowly like two immense disks that kept revealing new details to us, as if they were illustrating our conversation: we saw either the complete absence of architectural elements that would beautify the earth, or architecture in a state of neglect, or poorly conceived buildings in a scandalous state of dissonance with the surrounding natural environment. But our conversation gradually shifted to another subject.

"So what do you think?" I asked. "Could Esenin have visited Optina Monastery as a child?"

"I'm ninety percent certain that he did. His grandmother took him with her to all the monasteries in the region of Ryazan. And it's not far from here to there."

"And Bunin?"

"Probably not. . . . "

"But why not? It's so close. This is where he lived."

"I think that some evidence of it would exist somewhere: in his stories, his memoirs, his letters. No traces of a visit remain. I know that at the end of his life, in Paris, when he was desperately longing for Russia and was working on *The Liberation of Tolstoi*,[49] he mentally traversed the entire Kozelsk district and, it goes without saying, Optina. But in reality he was never there. If evidence is found somewhere in his archives, it will be a discovery."

"And how does one interpret the line by Anna Akhmatova, 'And I shall not see Optina anymore?'"[50]

"It's a mystery. It hasn't been verified in her biography that she visited Optina. But of course, she could have. . . . Heavens, did they take a count of who visited which monastery at which time? . . . And isn't Apukhtin's poem 'A Year at a Monastery'[51] of interest? Is it really about Optina? I'm embarrassed to admit that I haven't read it."

"You haven't read Apukhtin at all or just this poem?"

"I'm embarrassed to admit that I haven't read him at all."

"Well, how can I describe it? It's declamation. However, he was a friend of Tchaikovsky. An extensive correspondence between them exists. His poems were easily set to the music of romances.[52] They are somewhat pretty and sentimental, such as 'A Pair of Bay Horses,' or 'The Insane One.' You probably know 'Only Cornflowers, Cornflowers, / How Many of Them Gleamed in the Field. . . . '[53] He didn't like nihilists and destroyers:

> I hate to lie and be hypocritical,
> It's intolerable to live by denying . . .
> I want to believe in something,
> To love something with all my heart.[54]

"On the whole, we can find analogous poets even today. When we consider Apukhtin's monasterial poem, we know that he was, after all, from Kaluga Province and visited Optina many times. We can say with certainty that the poem formally describes Optina Monastery. There are signs of it in the description:

> The cells lie scattered about in a garden,
> Where multitudes of flowers and rare plants grow,
> (Our monastery had long been famed for its flowers)
> In spring the garden was paradise on earth . . .[55]

"But, strictly speaking, the poem describes a monastery in general terms. A man wants to flee from the turbulence of his passions, from his love for a woman, and so he hides in a monastery. But the woman and life emerge victorious. He runs to the woman. And that's the entire poem. Any monastery, not specifically Optina, could serve the purposes of the poem. And that's how it appears in the poem."

"And Leontiev?"

"Konstantin Nikolaevich?"

"Of course."

"So, what about him? He lived at Optina for many years. There's a whole volume of the letters he wrote from Optina, in particular to Vasily Vasilievich Rozanov. Moreover, the most interesting thing in it are Rozanov's commentaries to these letters. Of course, Leontiev was very conservative, but there's such thoughtfulness, such steadfastness in his positions. . . . For instance, he did not accept Dostoevsky's Christianity."

"And what did he suggest in its place?"

"Well, it's difficult to explain it in just a few words, one really has to read it."

"Both of them, Leontiev and Rozanov, are muddle-headed people."

"Yes, but their thoughts strayed into realms where you and I have never gone. . . . Such thoughts would never have come to us even in our dreams. . . . By the way, you wouldn't happen to remember what Blok wrote in one of his last articles about the synthetical character of Russian culture. . . ?"

At that moment I related the pertinent place in Blok's article to Volodya in my own words; but now, sitting at my desk, I stretch out my arm to the bookshelf and quote precisely:

> Russia is a young country, and her culture is synthetic. . . . Just as painting, music, prose, and poetry are inseparable in Russia, so philosophy, religion, community life, and even politics are inseparable, both from them and from each other. All together they indeed form a single mighty stream that carries within itself the precious burden of national culture. The word and the idea find expression in paint and an edifice; an ecclesiastical ritual is echoed in music; Glinka and Tchaikovsky bring to light "Ruslan" and "The Queen of Spades," Gogol and Dostoevsky describe the Russian elders and L. Leontiev, while Roerich and Remizov write about their ancient homeland. These are indicators of strength and youth. . . . [56]

Along the way we drove past many interesting places. Past Borovsk, for example, with its Pafnutius Monastery and the grave of the boyar Morozov's wife. We skirted around Kaluga, didn't stop in Peremyshl, didn't stop to see Tikhon Monastery or that same Polotnyanyi Zavod. Some of these places were located right by the road, while others lay off it; but we could have stopped at one place, then another, and we would not have regretted it.

But then, first of all, we wouldn't have reached Kozelsk on the same day and, second, we would have used up the capital of our attention. When you go to visit a place that is new to you, you have in reserve an unexchangeable "ruble of attention"—one hundred kopecks. And so, you can begin to spend these kopecks while still on the way there. Some of them will be left for the end of your journey, but then it will no longer be an unbroken ruble; instead, it will be one that has been exchanged. We entered Kozelsk and met Vasily Nikolaevich Sorokin, having in reserve a full-value one-ruble piece.

Vasily Nikolaevich was an elderly man with a shock of gray hair and a handsome and expressive but tired face. I would say that a kind of rebuke or reproach was fixed on that face, as if the man wanted to say to everyone: "People, what have you done? Why have you done this?"

This could have been the face of an artist or musician, or some kind of professor; but somehow this face seemed incongruous with a tiny museum of regional history in the tiny town of Kozelsk.

Incidentally, the museum can be called "tiny" only if one overlooks the fact that under its auspices are all the memorial objects in Kozelsk and its environs, including Optina Monastery itself.

We decided to save the next day, from morning until evening, for Optina; today we would see Kozelsk itself and toward evening would take a quick look at Beryozichi.

It's hard to imagine what Kozelsk looked like when its silhouette—that of a small town, almost a hamlet—was defined in the seventeenth century by forty churches. According to Vasily Nikolaevich's calculations, today there are six churches left, but only one of them helps to create the town's silhouette. To be sure, it's a very pretty one, and has an equally pretty bell tower. At least this one has remained intact, because if you take it away, cover it with your finger as on a picture, paste a piece of paper over it, then immediately the town of Kozelsk disappears, and what is left is an ordinary populated spot on a hill.

The other churches, despite Vasily Nikolaevich's assertions that they exist, in reality don't exist, except for one, a functioning one not visible from any part of the town. In the remaining cases there are just fragments, the lower parts of buildings, which are formless and adapted to other uses, such as a bakery.

I asked Vasily Nikolaevich to write down the names of all of the Kozelsk churches for me so that I could remember them. He wrote down these words: "In Kozelsk there are six churches, one of which is functioning. Two have remained intact. Two others consist of fragments (the Soshestvinskaya and Vozdvizhenskaya), while two, actually three (since the Nikolskaya church, though intact, has also been damaged), need to be restored: the Uspensky cathedral and the Pyatnitskaya church. The most valuable architecturally—

the Voznesenskaya church (1610) and the Soshestvinskaya church (1730)—lie in ruins, but they need to be and must be restored."

One of the ruins (the Soshestvinskaya?), the one near the museum of regional history, formed, along with its surroundings, literally a symbolic picture. To the novice's eye, of course. Later on, when we had met with the local officials and were engaged in an intimate conversation about the town and its external appearance, I mentioned this picture, and my interlocutors even gasped, "How could we not have noticed this ourselves!"

At once they made arrangements to set everything in order, and by morning orderliness prevailed. The picture I had seen was as follows: in a level, grassy little square stood a rectangular brick framework, as if roughly hewn from all sides, of a former building: it wasn't quite a barn, or a tower either. Most of all it resembled a transformer unit, only it was too large for that, and the way it was constructed of brick, crumbling all around and picked clean, made one think that it was not a transformer unit. Around this (I don't know what to call it), not unlike the engulfing waves of the ocean surf, a mass of empty boxes was piled up, like packing boxes with characteristic dividers creating cells in each box. They had probably contained mineral water.

"How is it that we didn't notice this earlier?" my interlocutors lamented. "What an eye! You should be a sniper. . . . "

"That's all right, I also like my profession. But someone ought to arrange for us writers to be issued milk because of the occupational hazards we endure."[57]

"What kind of hazards?"

"What do you mean? One gets tired of looking at such pictures . . . do you think it doesn't leave its mark?"

. . . We saw the Kozelsk train station. Right here Lev Tolstoi got off the train on his last journey and walked through this door into the one-story building of the station, which extended the length of the platform and retained the aura of the nineteenth century. On the left was the door into the waiting room into which they carried his things (for some reason it is locked now), during which time, right on this square by the station (now occupied by three trucks, four taxicabs, and a bus), he himself hired carriage drivers to take him to Optina Monastery. Right here on this square he scribbled his last autograph on a scrap of paper. The secondary-school pupil Tanya Tamanskaya had asked him for his autograph; she had recognized him while he was still in the train car.

At one spot in Kozelsk we went down to the river along a winding, cobbled descent. A long, narrow, little pedestrian bridge with railings had been erected across the river. But on the opposite shore we could see a road embankment of wooden paving-blocks, at which a genuine, high road bridge had once formed a junction. Across this bridge at one time, early in the morning, Alexander Sergeevich Pushkin had traveled, in the direction of Belev. He was on his way to Arzrum and took a detour in order to visit Ermolov.

We crossed the bridge to the other side and walked along the former road that lay between white willow trees. Some of them were very old and crooked. Perhaps one could describe them, too, as being decayed and in ruin. We encountered a man who looked about thirty years old. It was broad daylight, but for some reason he was unshaven.

"Would you happen to know when these white willows were planted?"

The man gave us a puzzled look, not understanding what we wanted from him, and suddenly he blurted out in a rather nasty tone of voice, "They should have been cut down long ago and sent to the Devil!"

Just like that! Why cut them down? Let's say that they are no longer useful, because they were planted as roadside trees and the road has long since been liquidated—but still, they aren't doing any harm. And why didn't the man think of planting new trees and making this path more orderly rather than cutting them down and sending them to the Devil?

. . . We stood for a while near the house in which Turgenev had stayed several times. It's still a good, solid house with a mezzanine. People live in this house, there are several families, but outside a memorial plaque has been attached to it. It stands on a quiet, almost provincial street. And it's named for the pioneer[58] Nikita Senin.

Should I describe how, on the following morning, we drove to the enormous and completely empty Kozelsk bazaar, where we hoped to buy something for breakfast, such as greens and sour cream; the modern concrete-and-glass, as if very original but actually mass-produced (some just like it probably exist in other towns like Kozelsk) House of Culture with a long-necked and long-legged flamingo in front; the hotel we lived in, which was rather cozy and quiet; or several merchants' houses that remained intact, which Vasily Nikolaevich showed us?

Sometimes as we were walking down the street he would say to his wife, Valentina Mikhailovna, something like "If I die, remember that in this house there is a stove with antique tiles. . . ."

. . . After this, according to our plan, we went to see Beryozichi. One has to travel several kilometers outside of Kozelsk through water-meadows, and then climb a hill to an old, shady park.

At one time the estate had belonged to Pushkin's uncle, Vasily Lvovich. I think it was like this: by inheritance Sergei Lvovich received Mikhailovskoe, while Vasily Lvovich acquired Beryozichi. It could have happened the other way around. At any rate, by 1877 Beryozichi was already owned by Dmitry Alexeevich Obolensky, the descendant of the Decembrist.

Having learned (in 1877) that Lev Nikolaevich Tolstoi was traveling to Optina, Obolensky sent horses to meet the train in Kaluga, and in this way was able to "engage" the famous guest immediately. It's true that Lev Nikolae-vich nevertheless spent the first night at Optina, but on the next day he indeed was the guest of the old prince. In the beautiful palace at Beryozichi, Tolstoi listened to the playing of Nikolai Rubinstein, and himself gave a reading of several excerpts from *Anna Karenina*, the novel he was working on at the time.

Sofya Andreevna wrote in her diary (quoting from Lev Nikolaevich?):
"It was very pleasant at Obolensky's: he has a kind wife, there were many young people, the host was very cordial, and the piano playing of Nikolai Rubinstein was delightful. All of this brought Lev Nikolaevich great pleasure."

As for the pleasure he himself gave to everyone by reading *Anna Karenina*, we won't attempt to surmise.

And here we are, too, walking about this same estate and park.

The fate of similar estates unfolded (if they remained intact) mostly accord-ing to the following scheme. As long as the estate (the palace and outbuild-ings) remained in good condition, a rest home or sanatorium would be situat-ed in it. Gradually the structure (along with the ponds and park) would lose its original well-groomed appearance and begin to look neglected, which appearance was not in keeping with that of a rest home or sanatorium.

At this point the estate would be given to the local state collective farm or to an MTS. This improved neither the external appearance nor the general condition of the architectural complex. As is well known, the MTSs were liq-uidated. On the premises of the estate one could find repair shops or some sort of schools for tractor drivers, agricultural schools, or special boarding schools. These were not isolated instances. In a village near my own, named Cherkutino, in a former nobleman's house (that of the Saltykovs), there is, coincidentally, a school like the one here in Beryozichi. It was established after the estate had been used as a sanatorium, and also as a center for a state collective farm. All according to the scheme.

It so happened that the director of the school came out to meet us, when we had just walked under the broad canopy of the park—more precisely, those trees that at one time had constituted a park but now only grew without form-ing a lane, or clearings, or picturesque stands.

The school was initially located in the palace itself, but several years ago there had been a fire. They managed to extinguish it, but still, it became tem-porarily impossible either to live or study in the palace. Meanwhile the local residents began to appropriate whatever they could from the place. What remains of the palace today is a mere box without roof or floors. All of this took place during the past few years. On his last trip here, Volodya Desyat-nikov had seen the palace intact and undamaged.

A new building was put up to house the boarding school, a long, one-story house of silicate brick.

"Well, what do you think, what should be done now with the former palace?" we asked the director of the school.

In a metallic-sounding voice, as if already used to defending his own opinion, the director proclaimed: "It should be torn down, and only torn down. So that not a trace of it remains."

One can understand the director's position: he lived under the constant threat of an event: once the palace's remains were torn down, life would be easier for him. Indeed, a huge, iron water tank weighing several tons protruded above the skeleton of the palace. At one time it had stood in the attic, but now it simply hung in the air and no one knew what held it there. Of course, if it were to fall (and it certainly would fall), and if children were to be under it. . . . So, was there no one who could remove it before this happened? No one in the entire region? In the entire province?

But meanwhile most of it was intact, solid, stable, and even beautiful: the foundation, walls, window openings, and ceilings between the stories. If people were to apply themselves and provide the necessary financial means, a wonderful palace would appear once again. After all, it would be cheaper than building a new one like this from scratch. And besides, who would build such a palace today anyway? And would our earth really be worse off if such a palace were to exist on it? . . .

We walked around the spacious box of the palace. We didn't hear the sounds of a piano, nor did we hear excerpts from the timeless novel. Glazed green tiling—remnants of the roof—lay here and there in piles. Bushes of lilacs, elderberries, and dog roses grew in all directions, disorderly and wild; they "stormed" the abandoned ruins. The belvedere that had once separated the palace, on the side of its facade, from the remainder of the park was lost, its bricks crumbling from the years and the rain. Not a single trace remained of the stone stairs leading to the park. We walked out to the edge of the park, to the edge of the hill, and before our eyes an incredibly vast valley opened up below, with a river and a lake, and distant little villages on other hills beyond this valley. How beautiful this place was, and still could be!

As he was saying good-bye to us, the director of the school, desiring to recruit us as potential allies in tearing down the palace, said: "My kids, they're very good kids. They are only a little bit . . . you know . . . only slightly. . . . "

"May God grant those kids of yours good health! But why should they determine the fate of a palace and its entire setting, the fate of a previous creation of human hands, a historical and cultural treasure that had existed for centuries before them and before all of us, too?"

. . . Optina Monastery itself also followed the aforementioned scheme. It started out as a rest home, then became a state collective farm, and now is an

SPTU, which stands for (if I'm not mistaken) "rural industrial-technical school." Between these points on the scale there were some other stages, but no one could tell us exactly what they were. Somehow the information escaped everyone's memory. But it is actually because of going through these nameless stages that the monasterial complex lost its entire architectural character.[59]

My friends Boris Petrovich and Tamara Petrovna Rozanov had lived near Optina at the beginning of the 1930s. Each of them has a separate set of impressions. Boris Petrovich used to go to Beryozichi to play billiards. At that time he was a young man, a boy. In Zhizdra they used to catch sterlet and huge, plump bream. The fish were caught by spearing them, but there was another method of fishing handed down from the monks; it was called fishing "with a net."[60] This was a very complicated method involving the screening off of a river with fir twigs, but it was described to me in detail.

Tamara Petrovna says that at that time everything at Optina was still intact (the buildings, that is), only in all parts of the rest home one noticed the strong odor of incense. They just couldn't air out the rooms enough to eliminate the smell. And all the graves near the cathedral were intact. Nothing monasterial was left inside the buildings and churches: everything had been thrown away; some items had been taken by local residents. A certain Zakharov, who participated in the confiscation of the monastery's property, hung some carpets around his house; but instead of carpets, they turned out to be horsecloths from the monastery's stable. It's rather amusing. Immediately after the monastery was liquidated, an auction was held right on the portico of the cathedral in order to sell such items as furniture, dishes, clothing. . . . Pafnutius's well, which was considered holy and to possess the power of healing but in reality was a deep, sulphurous spring, was blocked up with cement because people kept making pilgrimages to it for its water. But the spring burst out in a new place and found an outlet for itself right on the bank of the Zhizdra, several meters from the water, in a little ravine surrounded by dense bushes. . . .

This well and spring were the first things we saw at the former Optina, even before we had reached the monastery's wall. Now the well has become a square frame open to the sky, located on a square of about sixteen meters and filled with yellowish, standing dead (I would say) water with small pieces of garbage floating motionless on top—meaning that this water oozed out from below and collected on top of the cement gag. It was plainly noticeable that at one time excess water had continuously flowed out of the frame in a stream going in the direction of the river. On the path of this little stream depressions had been created where the water could collect and where people could wash their faces and hands and, if they desired, immerse themselves. There used to be a chapel above the frame.

After we had stood for awhile above the frame containing the "dead" water, we walked to the living spring that had forced its way to the surface of the soil in a new location. The water boiled up under a willow bush and rose in a mound, lifting sand of an unnatural gray color into its clear streams. We tasted the water, but it was so rich in various mineral salts that one could do only that—taste, but not drink it.

An elderly woman in a white scarf was sitting on a green bank by the extensive and still clear former monasterial ponds. My friends Volodya Desyatnikov and Vasily Nikolaevich walked on ahead to the monastery's wall, while I stopped near the woman and started a conversation with her. It turned out that she remembered the monastery before its liquidation and had seen its closing. . . .

"Did you have an interesting chat?" Volodya asked me after I had caught up with and rejoined my friends.

"It was fascinating. When I return to Moscow I'll certainly suggest a plan for us to be issued milk because of the occupational hazards . . . even a quarter-liter bottle for each of us. But for now give me a *validol*."[61]

. . . How can I describe in a word what Optina Monastery looks like now? There is a word—debris. The remains of the red brick wall look like they have eroded from the top or melted away. The monastery's towers, similarly, appear to have eroded or melted away. One cannot, however, describe the cupola of the Cathedral of the Entrance of the Holy Virgin into the Temple as having "melted away" from the top, for that would be an understatement. The cathedral has been decapitated, and without its cupolas the remains look like drums hanging in the sky. The Kazan church to the south of the cathedral has been decapitated in just the same way. The Vladimir church (once used as a hospital) has disappeared entirely. Melted away. Only the formless skeleton remains of the Church of St. Mary of Egypt, and it is overgrown with simple elder bushes.

Coincidentally, on this same day some Muscovites had come to Optina on an excursion. They came by bus. Mainly a group of sightseers from some NII,[62] along with several young women from the State Bank. Vasily Nikolaevich's wife, Valentina Mikhailovna—a tour guide of the Kozelsk Museum—escorted the Muscovites around the territory of the monastery, while Vasily Nikolaevich himself walked with us. But we all met by the graves, that is, near the Cathedral of the Entrance of the Holy Virgin into the Temple.

Is it possible to imagine what this place had looked like before its destruction? Yes, it's possible, because on occasion I've seen undisturbed monasterial graveyards elsewhere. Here, too, there probably lay gravestones with inscriptions, there probably were monuments, and there probably were flowers growing. It's difficult to imagine anything else: that this place with its pitiful fence could look solemn, though sad, and, just the reverse, that a solemn, though sad place could look as pitiful as this. But even for what remained here one had to thank Vasily Nikolaevich Sorokin, who had managed to move matters from a standstill, starting from zero.

When the specific opportunity to gather stones presented itself, not a single gravestone was left in its proper place. For some fantastic reason outside the bounds of common sense, they had all ended up scattered about the vast territory of the monastery. And this was not merely a handful of stones that one could throw all over the place; these stones were heavy, of marble or polished granite.

Later on, Vasily Nikolaevich gave me a memorandum consisting of a list of the people buried here. It reads very much like a memorial list. Well, we might as well remember these people:

   1. The Optina elders: Lev, Macarius, Ambrose, Anatoly, Varsonavry, and Anatoly.

   2. The Kireevsky brothers, Ivan Vasilievich and Peter Vasilievich.

   3. Natalya Petrovna, Ivan Vasilievich Kireevsky's wife.

   4. Alexandra Ilyinichna Osten-Saken, aunt and governess of Lev Tolstoi, who died in 1841.

   5. Nikolai Ivanovich Gartung, Pushkin's son-in-law, if he had lived to the time of his daughter's marriage; Gartung died in 1859.

   6. Major-General Andrei Andreevich Petrovsky, participant in the battles at Borodino and Leipzig, present at the capture of Paris.

   7. Countess Elizaveta Alekseevna Tolstaya, a relative of the Tolstois. Died in 1851.

   8. Relatives of Apukhtin, the Bilim-Kolosovskys: Varvara, Vasily, Nikolai, Matvei, and Dmitry.

   9. Grigory Voeikov.

   10. Maria Kavelina, née Nakhimova, sister of Admiral Nakhimov, hero of the Sevastopol War.

   11. Ivan Adamovich Pilissner, founder of the first institute of forestry in Russia, which was located at Optina. Died in 1815.

   12. Colonel Osip Osipovich Rosset, brother of the famous A. O. Rosset and a friend of Pushkin. Died in 1854.

   13. Colonel Alexei Osipovich Rosset, Pushkin's friend.

   14. Ivan Ivanovich Pisarev, father of the well-known critic. . . .

. . . We all crowded around the fence, beyond which lay only a few gravestones marking some graves. They were brought back here and placed at the initiative and by order of Vasily Nikolaevich, for no one knew anymore which gravestone should mark which grave. Vasily Nikolaevich measured the distance from the corner of the cathedral with his eyes, paced it out, and pointed at the earth: here lies Peter Vasilievich Kireevsky; here—Ivan Vasilievich; here, Natalya Petrovna. . . . I think that if his calculations were off by a step or half a step it would not be of great significance. They are all there, in the earth below. And, after all, it is we, the living, not they, who seek to remember them and who need the gravestones to mark their graves. Beyond the boundaries of the fence, off to one side, was a small mound, somehow not even the size of a real grave, with flowers on it. This marked the place (approximately) where Elder Ambrose was buried.

When we had walked over to the fence and for a time had joined the group of tourists from Moscow, Valentina Mikhailovna had just finished relating some facts to them: about the founding of the monastery, about the robber Opta, about the Optina elders, about Gogol, Dostoevsky, and Tolstoi, about *The Brothers Karamazov* and "Father Sergius," about the Kireevsky brothers, about Mme. Osten-Saken. . . . We could have left the graves and gone farther, to the skete, but at this point Vasily Nikolaevich assumed his wife's role as tour guide and decided to add some details to her account. He spoke for at least half an hour. It's a shame that no one had on hand the means to record the remarks; now I can recall only the main points, but without their nuances and liveliness, and moreover in my own interpretation.

"Remember this place, my young friends," Vasily Nikolaevich began. "You will probably live for a long time yet, and history will set before you, too, the question of humankind's relationship with the past. And if the question is not set before you, it will arise before your children. And from whom can children acquire intelligence and reason if not from you? Not from us. And we have acquired it . . . from those who lie beneath these gravestones.

"A person is a social, national, and historical phenomenon, and as such possesses three dimensions. People have a past, a present, and a future. Without one of these aspects they are not so much incomplete as simply nonexistent. They become like physiological entities—eating, drinking, and sleeping—but the social and national facets of their being do not exist, they have no history and, if you will, no country.

"In the East there is a saying: 'If you fire into the past from a pistol, the future will fire at you from a cannon.'

"A person capable of defacing a grave is also capable of insulting the living. A person capable of outraging his or her own mother will hardly desist from insulting or humiliating someone else's mother. Moreover, a person who destroys another's grave cannot be certain that his or her own grave will not be destroyed.

"The conception of one's homeland cannot be formed from speculative or philosophical ideas, from articles or scientific treatises. The meaning of the word 'homeland' is developed from concrete and visible objects: peasant houses, villages, rivers, songs, tales, and scenic and architectural beauty. 'Homeland' is not a notation on a map lying within the iron hoops of meridian lines, as one poet put it, but instead it may be a birch tree under one's window, to use the words of that same poet. One cannot love contours on a map or even the geographical map itself, but one can love a spring of water or a little path, a quiet lake and one's own house, friends and relatives, one's teachers and mentors. . . . Yes, love for one's homeland is formed from love for concrete things; if this love is illuminated and enriched by a love for traditions, legends, and history, then it will constitute 'culture.'

"A cowshed, steam engine, crane, and road roller are all indisputably useful, but one cannot love them. One can, however, love the Pokrov-on-the-Nerl Church, one can love the Cathedral of St. Basil the Blessed, the Moscow Kremlin, songs, the poems of Pushkin, and the novels of Tolstoi. Let us gather, bit by bit, pebble by pebble, everything that one can love and that in its entirety will constitute our love for Russia."

We walked for a little while longer across the wild, untamed grass of Optina, looking at the numerous, diverse, and unalike structures that didn't create any kind of ensemble, as well as the annexes clinging to the main skeleton, both outside and inside the monastery's wall. From Optina we went to the skete. The surrounding forest of mature pines that seemed made of copper remained completely untouched, to preserve appearances and order, for a skete had to be situated in a forest and the road to it had to lead through a forest.

The skete and everything in it had survived intact to a much greater degree of preservation than that of the main monastery. The half-kilometer's distance from Optina, the smaller main buildings and general solidity, especially of the outside wall (and of the other structures as well), indirectly had helped it to survive intact. It seemed as if a wave of destruction had passed over it. Similarly, a tornado can topple an oak tree and not touch a willow bush.

Although the main church of the skete—the Church of John the Precurser—also didn't have a cupola or a cross; although in the elder's house lives some sort of outsider who knocks down the walls in it and alters and reconstructs everything as he pleases (and there could have been a museum exposition here); and although one can no longer describe what exists here as a flower garden, still one has to admit that the skete remains in a happy state of preservation. The house in which Gogol stayed is whole and in good condition. And Dostoevsky's little house is also whole, and has even been restored. The building that housed the Optina library is whole. I repeat that, even though an outsider occupies it, the little house in which the elders lived is whole. Finally, the wall is whole. And the silence all around is whole, as are the pine trees that surround the skete.

As we walked through the skete I kept thinking about what I could take away with me from here as a keepsake: some sort of wooden scroll with remnants of gilding on it, or a little fragment of an iconostasis or an icon-case—sometimes a little fragment just happens to be lying around on the floor. . . . Even just an ancient iron nail, a small link of a chain, a shard from an icon-lamp (a little piece of stained glass), or a little corner of broken tile—one can pick up so many things in a place like this. It would lie on my desk, and I would know that this had come from the skete at Optina Monastery.

I actually did trip on something, and with an awkward movement of my foot dug out of the earth . . . but, alas! not something that had the distinctive marks of time and human hands on it, but a rather sizable (about the size of a man's fist), almost perfectly rectangular, flat, yellow, smooth flint.

Now it lies on my desk, and I use it as a paperweight so that papers don't fly in all directions when the room is aired.

. . . Shamordino was next. You'll recall that *Khadzhi-Murat* was written there. But the entire *nouvelle* was not penned there: Tolstoi worked on it for a very long time, and not less than ten versions separate the final variant from the original draft. But still, many pages of it were written at Shamordino; but most important of all is the possibility that in this very place, on the high and expansive bank of the Seryona River, the idea for this *nouvelle* was conceived. It's difficult to locate the appropriate line in the myriad volumes of Tolstoi scholarship, but the conviction exists that precisely in this place Lev Nikolaevich saw the thistle bush that suggested the idea of the tormented and tortured man who nevertheless did not give up: "I was returning home by way of the fields. It was the very middle of summer. . . . One can see a charming selection of flowers at this time of year. . . . I had gathered a large bouquet of various flowers and was walking home when I noticed . . . a marvelous crimson burdock in full bloom, of the type that in our region is called 'the Tatar.'. . . "

It turned out that right before driving into Shamordino we stopped at the side of the road in order to photograph the monastery's ruins from a short distance away.

"Why, here's the bush!" exclaimed Volodya Desyatnikov, having seen just a half-meter away from our car's tire a luxuriant bright-green and bright-crimson thistle bush in full July bloom. And in truth, I've had occasion to see many different kinds of flowers (I love to wander in fields and meadows), but had not yet seen one like this. Volodya set about photographing the bush from all sides.

"What do you think?" he began. "It's entirely possible that this is a descendant of that other, Tolstoian, thistle bush. How beautiful it is and how vigorous—what a fine creation of nature! . . . an ideal specimen. If we could only preserve its appearance, we ought to take it away with us and present it to the Tolstoi Museum. Moreover, the thistle bush is from Shamordino!"

The Shamordino Monastery was in all likelihood the youngest of all the monasteries in the Russian Empire. Since it was founded by Elder Ambrose, who died in 1891, it is clear that it could date only from the second half of the nineteenth century. It was called the Shamordino Women's Hermitage of Kazan.

The land on which the monastery stood, about three hundred desyatinas, had been donated by a female landowner, while the money for its construction came mainly from a very wealthy man, the tea merchant Sergei Vasilievich Perlov. The well-known tea store on Kirov Street in Moscow is located in the house he had owned. Of course, every Muscovite knows this house, which was built in a fanciful Chinese style. And thus was the Shamordino Women's Monastery built so quickly and splendidly with the money of the tea merchant Perlov.

But Elder Ambrose directed the entire project. Without his knowledge and approval not a single brick was laid at Shamordino. The elder created the Shamordino cloister (as we would say today) as a women's affiliate of the Optina men's monastery. The elder himself liked to reside here, if only because in this place he was bothered by fewer of the countless pilgrims, both male and female, who came from all over Russia. Besides this (I don't know whether this was of any significance), one cannot compare the location of the two monasteries. In general, I have not seen a place like Shamordino anywhere else. Of course, one can see hillocks, hills, slopes, and beautiful views, one can see meadows with a river meandering through them, but to see all of this arranged in such a remarkable way is rare and exceptional. At least, I haven't seen it.

A high, steep hill, on which one sees an oak grove and monastery, deeply concave, forms a horseshoe and with its sides embraces the smooth greenery of the meadows far below. In the distance, beyond the meadows and river, another hill echoes our own in form—it too stands in the shape of a horseshoe with its sides facing us and its deepest parts farthest away; it too is bright-green. But its verdure has already been dulled by a pale-blue haze: it has been softened and washed away. Of course, one can imagine such an ideal location as a model, but in order to visualize it in all its beauty, in its full scope, and in all its enchantment, one actually has to see it.

Elder Ambrose lived at Shamordino often and for long periods of time. A special shanty was built for him, if one can call a fine log cabin of seasoned, clear pine a "shanty." He even died in this house, and the nuns insisted that he be buried at Shamordino, but the church authorities nevertheless arranged for his remains to be transported to Optina.

I have already noted that without the elder's order not a single brick was laid at Shamordino. He even personally chose the site for Maria Nikolaevna Tolstaya's cell (again, if one can call a nice house a "cell"), and he made a sketch of it himself. However, I don't think that the Shamordino Monastery as a whole, as an architectural ensemble, was constructed according to the taste of Sergei Vasilievich Perlov. The synodal cathedral, burial vault, refectory, hospital, building for the nurses, house for the Perlovs, the "Case" for Ambrose's shanty—all of this was kept in a single, somewhat ornate and fanciful, red-bricked and ornamental-bricked style. It's not difficult to imagine it, because probably in every city (and sometimes even in villages—a church or chapel) one can see comparable monuments from the epoch of Alexander III,[63] when red brick without plaster or whitewash was utilized by architects as active material and was usually used in combination with wide and high white-latticed windows that were rounded at the top. And whether there was a greater or lesser number of little peaked towers, whether the style was more ornate or more severe, may in fact have depended on the taste of the tea merchant Perlov.

And, for all this, the main source of Shamordino's popularity, especially in our time, was neither its delightful location, nor Perlov the merchant, nor Elder Ambrose, but the fact that for twenty-one years Lev Tolstoi's sister Maria Nikolaevna had lived there, and that Tolstoi had often visited her there. Having left Yasnaya Polyana and spent the night at Optina, Tolstoi quickly went to Shamordino, where he hoped to remain. He even tried to rent a little place in the village and put a deposit on the place—three rubles.

One can read in detail about Maria Nikolaevna, her life, her unsuccessful marriage, her children, her relationship with her famous brother, their correspondence and conversations, their tender friendship that grew stronger over the years, and their old age, in the many memoirs, notes, diaries, and letters, including (I recommend) the reminiscences of Maria Nikolaevna's daughter, Elizaveta Valeryanovna Obolenskaya.

Incidentally, the latter happened to be at Shamordino visiting her mother when they were struck, as by an avalanche of snow, by the news that Tolstoi had left Yasnaya Polyana. Elizaveta Valeryanovna recalls:

> On the day of October 29 Mother and I went out for a walk. Since the weather was cold and it was very muddy, we didn't venture beyond the fence. We met a nun who had just returned from Optina Monastery. She told us that she had seen Lev Nikolaevich there, and when he learned that she was from Shamordino and was returning there, said, "Tell my sister that today I'm coming to see her."
>
> We were very agitated by this news. The fact that he had decided to leave in

this kind of weather did not portend anything good. We hurried back the way we had come and began to wait for him and guess what his arrival could signify. We waited for a long time; finally he arrived at our place at six o'clock, when it was already completely dark, and he seemed so pitiful and old to me. His head was wrapped in a brown hood, out of which his gray beard emerged rather pitifully. The nun who had escorted him from the hostel told us later that he had walked to our place with unsteady steps. Mother greeted him with the words, "I'm happy to see you, Lyovochka,[64] but in such weather! I fear that something is wrong at home."

"Everything is terrible at home," he said and began to cry. . . . He said that he was thinking of staying for a while at Shamordino.[65]

Makovetsky describes that same evening in these words:

When I came in, they were engaged in pleasant, calm, and lively conversation. . . . The talk they had was one of the kindest and most charming that I have witnessed during the last five years. L.N. and Maria Nikolaevna were happy to see each other, felt joyful, and had already calmed down. The excitement had passed, and what remained was a quiet heart-to-heart talk, intertwined with reminiscences and humor, the last of which Maria Nikolaevna possessed to no lesser measure than L.N. She was an amazing storyteller. Brother and sister had not seen each other since the summer of 1909.[66]

And the evening of the next day found them once again conversing peacefully. As he was leaving, he told his sister that they would see each other again in the morning.

However, that evening Alexandra Lvovna arrived. What they talked about remains unknown, but early the next morning, around five o'clock, Tolstoi rushed away without even saying good-bye to Maria Nikolaevna. The rest is well known: his illness, Astapovo, and his death. . . .

And here we are, standing on the spot where at one time Maria Nikolaevna's little house had stood, where brother and sister had seen each other for the last time, and where Tolstoi had spent the last two peaceful, cozy, cheery, warmhearted, and emotionally satisfying evenings of his life.

The little house stood on the very edge of a hill, and right from it the surrounding land descends sharply, but not as some kind of precipice, or as an unexpected cliff, or as bare clay soil or a rocky drop, but as a green (and in summer still covered with flowers) mountain slope. There is much land and sky to be found far and wide—Father Ambrose chose a pretty good spot for the convent. And indeed, a nun is a nun, but a countess is something else. And not just a countess, but a Tolstoi. Father Ambrose's convent pleased her.

The house was taken apart and moved to Kozelsk. People say that even today it remains intact. And the address is well known. Besides this, a house exactly like it, a replica, is located right here in Shamordino near the former hostel. So that, at worst, they could even move it in order to restore this wonderful, memorable place to its former appearance. And even Kozelsk is located, not beyond the mountains, but only some fourteen kilometers away.

From the spot where Maria Nikolaevna had lived we went, completely logically, to her resting place after her death. Here things were more complicated. . . .

It is known for certain that her grave had been near a linden tree, and this tree was also well known. But a family had been given this plot of land, along with the tree, in order to build a house, and the house was built. The house is a common country *izba*, a typical peasant homestead with the usual yard and vegetable garden, and also a small external yard for household purposes. In this yard (I'm ready to take any oath you please) there is a dog kennel, a pile of logs, and a heap of manure. Right in the middle of all this stands the old linden tree. It is intact, which one can't say about the burial mound—it has gradually been leveled down. Only a few people, like Vasily Nikolaevich (and now including myself and Desyatnikov), know that in a small household yard, at a depth of three *arshins*,[67] lies the beloved sister of Lev Tolstoi.

A sharp-witted man removed the headstone from the grave and took it away. He chiseled an inscription into the stone and placed it on the grave of his "old lady" in a country cemetery of a village that is well known. It wouldn't be difficult to find.

Vasily Nikolaevich Sorokin said, "I keep wondering whether I should dig up the grave and transfer the remains of Maria Nikolaevna to another place."

Whether right or not, we heatedly began to dissuade him from taking this action. It's not that the remains need to be moved, but rather that a dignified, or at least decent, appearance needs to be restored to this place. Can it really be easier to transfer a person's remains, to turn her bones upside down, than to move an *izba* from one location to another? Can it be that our country has so little land? And aside from this, they relocated Gogol, and Aksakov, and Derzhavin. . . . Isn't that enough?

In the case of many large and solid monasterial structures, they are adapted somehow to fill various needs. An agricultural school is housed (according to our scheme) in the main cathedral. They created two floors inside the cathedral by adding a ceiling, but still it's worth noting that, although many different establishments find shelter in similar buildings, they are nevertheless, aside from everything else, not suited for other purposes. They are unsuitable either for workshops or for various offices. They are not even suited for storehouses.[68] Children should attend school in a modern and well-lighted building with spacious classrooms that are warm and dry, not in an enormous cathedral whose extensive roof leaks, of course, in many places, gradually softening and eroding the brick.

We also stopped in for a few minutes at the "Case," in which Elder Ambrose's shanty had once been preserved. It was a solid, brick building large enough to contain a high and ridge-roofed *izba*, and moreover to allow one to walk completely around this *izba*, and even to allow (indoor) flowers to grow all around the *izba*. Later on, tractors and trucks were repaired here. Everything smelled of gasoline.

The house's fate is well known. In the 1920s in a nearby little village (the same one in which Lev Tolstoi tried to rent a place) there was a fire. Several *izbas* burned down. One family that had lost all its possessions relocated Elder Ambrose's shanty to the place where its house had stood before it burned down. We took a short trip to see it. It's in excellent condition. Inside we were struck by the height of the ceilings, so incompatible with, and so unexpected in, a peasant's *izba*.

"You know, such an *izba* isn't suited to our needs," complained the hostess, an elderly woman—an old woman, one might say. "It's not adapted for our household. The ceiling is so high, just try to brush away a spider web, you won't reach it, so you have to get a long pole. And we don't need such a large living space, we have to heat it, you know. Well, this is what we got, we can't do anything about it."

While we were walking to this village, Volodya Desyatnikov kept "inciting" Vasily Nikolaevich to buy back from the old woman for the Kozelsk Museum the painted wooden cross that had belonged to Elder Ambrose and had been acquired by the new owners along with the house.

"Vasily Nikolaevich, I'm telling you that the cross will get lost. Either they'll throw it away after the old woman dies, or swindlers-black marketeers will beat you to it and buy it or deceive them somehow, or else even steal it. Before it's too late, take it with you, give her a five-ruble note, register it. Our descendants will thank you."

Vasily Nikolaevich made a decision and, it seems, even had five rubles on him.

The hostess showed us the cross, acting as if she attached no importance to it at all. Without a doubt the cross was the one in question, which had been described more than once in various inventories, books, and even appeared in several photographs depicting the furnishings of Ambrose's shanty. For example, this cross is distinctly visible in the photograph in which the elder is lying on the sofa while the cross hangs on the wall over his head.

The old woman's words about the house, that it was not really comfortable and they didn't need one like it, and moreover the indifference with which she showed us the cross and even handed it to us so that we could hold it, gave the art historians some hope. As if speaking about a matter that had already been decided, Desyatnikov said: "Vasily Nikolaevich is the director of a museum. Why don't you sell this cross to him? . . . "

At this moment I saw a momentary transformation in the old woman: nothing seemed to have changed, but in the features of her face, in her eyes, and in the inflection of her voice immediately appeared something immovable, iron-willed—qualities with which people in the past ascended a bonfire while holding up two fingers in the air.[69]

"No, I don't think so," was all she said.

. . . Our car rushed back to Moscow. We talked about what we had seen (and we saw a hundred times more than what I describe here), and about what we should do with our newly acquired knowledge.

"I think you're right," Desyatnikov agreed. "One ought to approach this problem broadly and comprehensively. Indeed. Tourists will travel a great distance to see Optina (when they restore it), so why not let them see Beryozichi too, at the same time, as well as Shamordino Monastery and the antiquities of Kozelsk. . . . But you know what the main problem is here? Not so much the preservation of monuments, but rather how they are utilized. All right, let's restore all one thousand Russian monasteries, and the tens of thousands of churches and country estates, but what shall be done with them afterward? How should they be utilized? Should we house the same RPTUs in them? Museums? Storehouses? Hotels? Exhibitions? Artists' houses? This problem has to be settled before anything else is done.

"No, first of all we must restore and preserve them. The money will be recouped in tourists' fees. Beauty doesn't necessarily require a utilitarian application. I'll even repeat it one more time. In order for a generally benevolent atmosphere, a generally favorable climate, to develop around monuments of antiquity, an authoritative, comprehensive decision on the restoration of all of them is needed.

"In the final analysis, this is the beauty of our earth, this is our culture, this is the measure of how civilized we are. . . . "

. . . "What kind of little poem," Volodya began after a while, "did Vasily Nikolaevich give you when you were saying good-bye? Could it be his own?"

"I haven't read it yet. Here, take a look at it." I took out of my pocket a small piece of notebook paper folded in half and handed it to Volodya.

"Someone named Tatyana Alexandrovna Aksakova from Leningrad wrote it, evidently an old woman," Volodya said, after glancing at the bottom of what was written and quickly scanning the lines. "It's about Dostoevsky, and Optina. . . . "

"Read it aloud." I held on to the steering wheel and looked ahead while Volodya read:

> He grew fond of the village environs,
> Loved the vast expanses of those places
> Where near the city of Kozelsk stands
> The famous Optina Monastery.
>
> Entire generations of Russians
> Considered this land sacred;
> In search of revelations from on high,
> Gogol, Dostoevsky, and Tolstoi came here.
>
> And in spring the sticky green leaves[70]
> Blossomed with such joy and appeal,
> And the fish splashed by the ferry,
> And the dog-rose bushes bloomed.
>
> And the evening sky, sparkling like diamonds,
> Said to the calm surface of the waters:
> "Surely Alyosha Karamazov
> Will pass by on the dewy path."

CIVILIZATION AND LANDSCAPE

FIRST OF ALL, I must admit that although my presentation is called a paper, I'm still not a scientist or a researcher, and consequently what I say will scarcely have any scientific value. At best, I'm a poet capable of perceiving the beauty of the world around me in a distinctive way; at worst, I'm a publicist who occasionally strives to examine a particular issue in a journal article.

Because of the scientific nature of this paper it seems appropriate to clarify its terminology: what is "civilization" and what is "landscape"? Both the one and the other possess a multitude of different meanings. Let's take the most elementary sources—encyclopedic dictionaries. In an old Russian encyclopedia entitled *Brokgauz and Efron*,[1] a long and detailed article is devoted to "civilization." In it, civilization is defined as "the state of a people, which is attained through the development of its public opinion and social life, and which is characterized by its movement away from its original beauty and wildness, the improvement of its material conditions and social relations, and the advanced development of its spiritual life."

The word closest in meaning to that of civilization is "culture." However, Wilhelm Humboldt distinguishes between civilization and culture, while Buckle understands civilization to refer predominantly to the spiritual side of life: "A dual movement, moral and intellectual, comprises the very idea of civilization and contains in itself the entire theory of spiritual development."

Further, one encounters the definitions of "civilization" provided by Guizot and Klemm, Fourier and Morgan, Taylor and Lippert, Hegel and Kant, Tolstoi and Gottenroth.

The *Great Soviet Encyclopedia* remains silent about the spiritual aspect of civilization and is satisfied to define the term in merely a few lines, if one discounts an extensive citation from Engels. After several lines it becomes apparent that civilization is "a level of social development and *material* culture [emphasis mine—V.S.] attained by means of one or another kind of socio-economic formation."

And Oswald Spengler, for example, considers civilization to be the final stage in the development of a given culture, its old age, its end.

Concerning "landscape," the encyclopedic dictionaries sidestep this concept and describe it solely as a genre in the art of painting. However, in one of his articles Jean Zeitoun offers several definitions of "landscape," in which it becomes manifest that landscape has meaning in the following contexts: naturalistic, geographic, biological, ecological, psychological, social, economic, philosophical, aesthetic. . . . Zeitoun specifically states, "as we examine the diversity of meanings contained in the word 'landscape,' and the diverse applications of this word in various fields of thought, we reach the conclusion that the concept 'landscape' possesses a rather indefinite semantic meaning. . . ."[2]

It's true that the author of the article ultimately arrives at a formula which states that landscape must be examined as the product of the interaction between the individual and the surrounding environment, in accordance with a specific view of this environment.

But this is only one of the formulas, and it reminds me of a well-known moment in human history:

In the nineteenth century, while French and German philosophers, as philosophers in general, were struggling to define the concept of freedom (in the philosophical, moral, social, and psychological senses), Bulgarians, for example, simply needed freedom—a freedom as precious as bread, as precious as air, and as precious as those things necessary for life. And it turned out that for the Bulgarians of that time what was more important was not for philosophers to define and formulate the notion of freedom, but rather for Alexander II to advance the Russian troops and bring them freedom on bayonets, on those proven weapons of coercion and violence.[3]

Similarly, what I'm going to say here has more of an emotional, rather than a purely scientific character.

For the next half-hour let's agree to consider "civilization" as the contemporary (twentieth-century) stage of the development of humankind, while "landscape" will constitute what people who inhabit the earth and can turn in all directions see all around them.

Earth is a cosmic body, and we are all none other than cosmonauts making an extended orbit around the sun and together with the sun through the boundless universe. The life-support system on our marvelous spaceship is designed so ingeniously that it constantly renews itself, and in this manner ensures the possibility for billions of passengers to travel over the course of millions of years.

It's difficult to imagine cosmonauts flying in a spaceship through cosmic expanses and consciously destroying the complex and delicate life-support system calculated to last the entire trip.

Yet, gradually, though consistently, with startling irresponsibility, we are disabling this life-support system, poisoning rivers, cutting down forests, ruining the World Ocean.

If the cosmonauts on this small spaceship were to start cutting the wires, unscrewing the screws, and boring little holes in the plating, then these acts would have to be considered suicidal. And, in principle, there is no difference between a small spaceship and a large one. It's a question only of proportion and time.

From a pessimistic point of view, humankind can be considered a disease peculiar to our planet. The microscopic (in planetary and even more so in universal proportions) beings established themselves, are multiplying and swarming. They gather in one place, and immediately there appear deep ulcers and various growths on the earth.

It takes only the introduction of a drop of harmful (from the perspective of land and nature) culture into the green mantle of the forest (a brigade of lumberjacks, a barrack, a couple of tractors), and immediately from this place the characteristic, symptomatic, diseased spot begins to spread.

They scurry about, multiply, carry on their affairs, eating out the bowels of the earth, depleting the fertility of the soil, polluting its rivers and oceans, the earth's very atmosphere, with their poisonous refuse.

It's unclear what this peculiar illness called "humankind" portends for our planet. Will Earth have enough time to develop an antidote before it is too late?

Hundreds of books have been written that contain statistics, calculations, prognoses, and recommendations concerning the disruption of the ecological and chemical balance on our planet. But this is not the topic that concerns me at the moment.

Unfortunately, just as vulnerable as the biosphere, just as defenseless against the pressures of so-called technological advancement are such concepts as silence, the possibility of solitude, and consequently the personal, one-on-one, intimate, I would say, communion of the individual with nature, with the beauty of our earth.

On the one hand, the individual who is oppressed by the inhuman pace of contemporary life, its congestion, and its huge stream of artificial information, loses the habit of communing spiritually with the external world; on the other

hand, this external world itself has been reduced to such a state that it sometimes doesn't invite this individual to commune spiritually with it.

In his remarkable article "Urbanisation et Nature," Yves Betolaud (of
France) rightly states:

> The gradual degradation of the environment produces psychic and physical dis-
> orders. . . . Because of the inhumaneness of cities, people are transformed into
> puppets that are jostled and herded like cattle along the corridors of the metro. . . .
> It should be noted that since the beginning of this century beauty has been absent
> from the development of our cities, leading to the degradation of the world
> around us: urban environments no longer satisfy our needs, while in agricultural
> regions what meets the eye are landscapes that have been reduced to total chaos.[4]

Thus, in the problem "civilization and landscape" we can clearly identify
two aspects: first, the objective degradation of the landscape; and, second, the
changes in the individual's perceiving apparatus.

Humankind is currently preoccupied with what I would call a "bacchanalia
of accessibility." I can assure you that the snow-covered mountain in the Caucasus which glistens in its inaccessible heights and which Lermontov fancied
as the footstool of God's throne wouldn't be what it is if we could get there in
ten minutes by ski-lift. The atoll that can be reached after several months on
a sailing vessel would look different to us if we could fly up to it in a helicopter,
on a flight that lasted from breakfast to lunch. This bacchanalia of accessibility permeates the entire register of our communion with the external world,
from the mystery of a flower to the mystery of the moon, from a woman's love
to lightning and thunder. I only fear that in the midst of universal accessibility a concept such as beauty will gradually become inaccessible to us.

As birds have been created to fly and fish to live in water, so people have
been created to live with nature and constantly commune with her. Ancient
Indian wisdom posits that, for his or her spiritual and physical well-being, an
individual should look at the verdure of the earth and at running water as much
as possible.

And people have lived in the lap of nature and constantly communed with
her from the very beginning, and right from the beginning there existed two
aspects in a person's relationship with his or her natural surroundings: *usefulness*
and *beauty*. Nature fed, satisfied, and clothed the human being, but at the same
time, with her unsettling, divine beauty, she always exerted an effect on his or
her soul, creating feelings of wonder, veneration, and rapture.

Let's read half a page of Tolstoi's prose. Take the *nouvelle Kazaki* (The Cos-
sacks, 1863). Olenin is traveling by post chaise from Moscow to the Caucasus:

... No matter how hard he tried, he couldn't find anything beautiful in the view of the mountains about which he had read and heard so much. It occurred to him, that ... the particular beauty of snow-covered mountains ... was as much a fiction as the music of Bach and *love* for a woman, neither of which he believed in—and he stopped anticipating the mountains. But on the next day, early in the morning, he awoke in his chaise from the crispness of the air and looked indifferently to his right. It was a bright and clear morning. Suddenly he saw, at about twenty paces away, as it seemed to him at first, pure white masses with their delicate outlines and the fantastic, distinct, airy line of their peaks and of the distant sky. And when he comprehended ... the entire limitlessness of this beauty, he became afraid that it was an apparition, a dream. He shook himself in order to wake up. But the mountains were still there.

... From this moment on, everything that ... he felt took on the new, sternly majestic character of the new mountains. All of his memories of Moscow, the shame and regret, all the banal dreams of the Caucasus, all this vanished and did not return anymore.[5]

And thus, the beauty of the mountains dispelled all that was petty in his soul, giving life to new and positive forces, ennobling the man. At the sight of the stately snow-covered mountains, the deep blue of the heavens, the expanse of the sea, a silent grove of birches, or a maturing cornfield, suddenly all that has welled up in the soul of an individual consumed by his worries, all that is petty, vain, and fleeting, melts away. The soul touches what is exalted and eternal. "And in the heavens I see God," the poet Lermontov wrote about such a moment.[6]

The old, gnarled oak, which had become green once again and was covered with thick, succulent foliage, revealed to Andrei Bolkonsky the mysteries of an undying thirst for life and eternal rebirth, by means of that particular simplicity and certainty unattainable in a philosophical treatise. The deep starry blue of the nocturnal sky returned to Alyosha Karamazov the hope that had almost given way in his soul, while the majestic images of the vast steppes deeply imprinted themselves on the soul of the little hero of Chekhov's steppe, and moved him.[7]

This, of course, illustrates the level of perception of the beauty of the world by highly educated, cultured, and, more precisely, civilized persons: Tolstoi, Dostoevsky, and Chekhov. But it is indisputable that the beauty of nature has always acted upon individuals and focused their consciousness, enlightened them, and made them kinder, better, richer.

Landscape is synonomous with beauty, and beauty is a spiritual category. Not by accident has landscape become the subject of art, the subject of painting, literature, and even music. But landscape first became the subject of painting. Although landscape has always played a supporting role in literature, in painting it developed as a genre and assumed a significance all its own.

As early as the period of red-figured vase painting in ancient Greece we find examples of keen insight into nature.[8] As a consequence, through the Middle Ages, through the epoch of the Renaissance, through Giorgione and Giotto, through El Greco and Dürer, through Brueghel and Poussin, Ruysdael and Dughet, through the great artists of Holland, through the painting of France, Germany, Spain, and England, through the painting of dozens of artists, humankind has advanced toward a penetrating and spiritual understanding of the surrounding world of nature, since landscape painting has always considered as its highest calling the reproduction not merely of scenes of nature as such, but rather of those sensations, feelings, and movements of the soul that surface when a person contemplates his or her surrounding environment.

One should give nature its due, namely that contemplating her beauty arouses in a person's soul the most elevated, pure, and radiant feelings and lofty thoughts—and herein lies that precious, priceless quality inherent in nature.

A feeling for one's native surroundings has always entered and still enters into a concept as important as love for one's homeland, along with a feeling for the history of one's own country and its people. People everywhere cherish and hold dear their native environment; but if we ponder this notion we may conclude that this appreciation for our native environment is not spontaneous. In apprehending nature, we involuntarily stir up emotional reserves accumulated from reading our poets and writers, contemplating paintings, and listening to music. In other words, our very sensitivity to nature is organized, learned, and traditional—in short, culturally determined.

Each nationality has had and still has its own poets of nature who are sometimes little known to the people of other countries. But if the earth is our common homeland and if, little by little, we develop an appreciation for this common homeland—before the formidable countenance of the universe—if we acknowledge that perhaps there exists nothing in the universe more beautiful than our planet (and our earth really is wonderful!), then for these thoughts, too, we are indebted to our enlighteners: to the artists and poets of universal significance who have helped us to understand beauty and have facilitated our love for it.

It's easy to point out that, in creating their landscape paintings, artists of all time periods and nationalities have almost never depicted nature on their canvases without signs of human activity. A little bridge here, a small chapel there, a church, section of fence, castle, tiny village, boat, horseman, path, road, lighthouse, sailing vessel, cows grazing, a plowed field, a ripening field, garden, windmill, water mill, puffs of smoke above the rooftops. . . . Every imaginable sign of human activity.

This can be explained partly by the fact that we rarely see the earth in a state untouched by human activity. But the main thing is that up to a certain point such activity enlivened and improved the earth's beauty, introducing those brushstrokes and details that made this beauty more inspiring, closer to our hearts and—if one does not fear such a wholly unscientific word—dearer.

Signs of human activity did not corrupt nature, initially because people could not totally ruin her owing to her vast proportions, and subsequently— up to a certain point—because people thought about and took care of the beauty of the place where they lived, i.e., were concerned about the beauty of their surroundings. Up to a certain limit, all those changes that people have wrought in the world were harmoniously incorporated into the earth's land- scapes. A ripening field of grain, a little bridge over a stream, a herd of cattle grazing, or small one-story houses surrounded by trees in and of themselves could not destroy the enchantment of the earth, especially if we keep in mind that people consciously strove toward beauty.

But at some point people went beyond the limits, and they did so in two ways.

First, the scope of human activity changed. This is the objective reason.

People could not yet, did not yet know how to create Cyclopean waste heaps, or to build monstrous dams, or to point countless factory smokestacks toward the sky, or to excavate deep quarries, or to cover the earth's large flat surfaces with the concrete formations of cities, or to entangle the earth in modern superhighways, or to create artificial reservoirs that inundate vast expanses of the earth with a shallow and murky layer of water, as, for example, the flooding of the entire fertile meadowlands—the milk-rich and nectarifer- ous floodlands of the Volga—as the flooding of the unique Tsymlyansk vine- yards. And if the change in the direction of Siberian rivers toward central Asia is carried out, then in the lower reaches of the rivers it will be necessary to flood areas equal in size to that of an average European country.

Naturally, human activity of such proportions cannot avoid fundamentally altering the external appearance of the earth. Well known is the tragedy of Lake Sevan, that high-mountain pearl in Armenia, that precious pale-blue stone in the attire of our planet. An opening was bored into the mountain— and water started to flow out of the lake. It's true that the water's action rotates turbines and generates electricity (which could have been generated by other means), but the lake grows shallow, the water level in it drops sharply, its banks dry out or become swampy, the island turns into a peninsula, and springs of clean water dry up in the surrounding areas. The majestic is transformed into the pitiful.

But we cannot blame everything on the scope and range of human activity.[9]

After all, the Egyptian pyramids in their dimensions exceed those of many contemporary structures, but can we maintain that they have disfigured the earth's landscape, that, in looking at them, we experience unpleasant or oppressive feelings, that is, what are considered to be negative emotions? On the contrary, do they not represent one of the rarest and most amazing adornments of our planet?

We can object that the perception of landscape is a matter of taste. One person likes the Taj Mahal or the Cologne Cathedral, while another—the skyscrapers of New York. But still, there does exist an objective division of things into what is beautiful and what is not, into what evokes rapture and what evokes revulsion.

One can look at a star or a squashed frog, at a glade in bloom or a garbage dump, at a clear stream flowing over pebbles or a gutter, at the latticed gate to the Summer Palace or the chain-link fence of an auto depot.

Of course, the sight of a cemetery is inherently sad, but if this cemetery has been neglected, made filthy, or purposely defaced, then its appearance becomes not simply sorrowful, but intolerable.

Beauty lives in a person's soul and engenders his or her physical need for it, akin to that for eating and drinking.

The chain-link fence of an auto depot cannot be considered as beautiful as the latticed gate to the Summer Palace, for it has a different function, and the law of function must be acknowledged. But we can agree that even this same fence may be beautiful in its own way or ugly in its own way, depending upon the sensibilities of the people who erected it; or, in other words, did the idea of beauty live in their souls, were they aware of the need for beauty, and did they transform this need into reality?

I haven't really traveled the world that much, but still I have seen France, England, Germany, Denmark, Czechoslovakia, Poland, Bulgaria, Hungary, China, Vietnam, and Albania. . . . Moreover, I have traveled in many parts of the Soviet Union. In making observations, juxtapositions, and comparisons, I can state that often the most modern, most industrial and grandiose structures nevertheless are beautiful in their own way, and are even elegant. One cannot say that these structures have imprinted themselves harmoniously upon the landscape, since they themselves define it and become the landscape; but still, one cannot say that they are disgusting or ugly either.

The majestic doesn't necessarily have to be enormous. In order for a swan to be majestic, it doesn't have to be the size of an elephant. A two-story building, such as the Mikhailovskoe Palace in Leningrad, may appear more stately than a one-hundred-story box. But the reverse principle also exists: not everything that is grandiose and enormous is necessarily ugly. Our earth is large enough to "assimilate" and adapt to rather large structures. After all, Everest,

Fujiyama, Elbrus, Mont Blanc, and Kilimanjaro do not mar the external appearance of our planet.

However, people preoccupied with only economic or political considerations may lack one simple criterion: "How will it look?" How will it look today and, moreover, how will it look tomorrow?

In his book *Ob arkhitekture* (On architecture),[10] the Soviet architect Andrei Konstantinovich Burov matter-of-factly makes the following significant statement: "First of all, in a short period of time one should construct high-quality dwellings without ruining the face of a country for centuries in the process."

A wonderful and ominous sentence. Wonderful in its concern for the face of a country but ominous in its recognition that the face of a country *can* be ruined, and not for a year, not for two, but for entire centuries. The main thing that follows from Burov's statement is that there really does exist such a concept as "the face of a country," and that it is defined not only by geographic setting (a mountainous country, flat country, forested country, etc.), but also, and not to a lesser degree, by human activity.

Just as an artist creates a landscape painting, so an entire people gradually, even unwittingly, maybe brushstroke by brushstroke in the course of centuries, creates the scenery and landscape of its country.

Hills and rivers, trees and flowers may look the same in two different countries, but the brushstrokes introduced by a person (by a people) ultimately create this or that characteristic picture—and consequently the face of Germany differs from that of neighboring France or of neighboring Poland, from the face of central Russia or the Ukraine, while the face of Japan differs from that of the island of Sakhalin, which it resembles geographically.

The face of old, prerevolutionary Russia, for example, was defined largely by the hundreds of thousands of churches and bell towers situated throughout the country, primarily on elevated locations, and which defined the silhouette of each city (from the largest to the smallest), and similarly by the hundreds of monasteries and countless numbers of windmills and water mills.[11] Not an insignificant portion of the setting and landscape of the country was contributed by the tens of thousands of country estates with their houses, parks, and systems of ponds. But, of course, most prominent were the little villages and hamlets with their white willows, wells, sheds, little bathhouses, paths, yards, vegetable gardens, fallow land, fences, ornamental window carvings, roof ridges, little porches, markets, sarafans, round dances, haymaking, shepherds' horns, sickles, flails, costumes, thatched roofs, small private fields, horses ploughing. . . . One can imagine how radically the face of the country changed when all these factors that define landscape disappeared from the face of the earth or altered their appearance.

Just as landscape painters put their soul into their work and create a landscape, practically speaking, in their own image, so the soul of a people and that ideal of beauty which lives in the soul of this or that people are assimilated harmoniously into the scenery of a given country.

But of course, it is a bad thing if the soul is asleep, if it is distracted, absorbed in side-issues, concerns, noise, self-interest, or other considerations; it is even worse if the soul is dead or, to be kinder and more precise, in a state of lethargy. At such times spirituality also disappears even from the landscape. Scenery is scenery, but it seems emptier, its form remains in the absence of content, it exudes coldness, alienation, indifference, and hence emptiness. What ceases to matter to the isolated individual and an entire nation is: How will it look? How will the house, village, river, valley, hills, the country as a whole look? What will be the face of the country?

There are departments for fieldwork and the extraction of useful fossils, for the building of roads, for agriculture, for electrification, and for the light, heavy, and automobile industries, but there is no department for the external appearance of the country (of the earth), for its tidiness, orderliness, spirituality. . . . We think about the structural integrity of buildings, about the character and magnitude of earthwork, about the quantity of wood, about centners and tons, about cubic meters and square meters, but we don't consider the question: How will it look? How will it look, not only in and of itself, but in combination with the surrounding environment, with the countryside, in harmony with local traditions and projections for the future?

Landscape in all its complexity and totality embraces not only the face of the earth, the face of a country, but also the face of a given society. A forest cluttered with garbage, worn-out roads with abandoned cars, rivers that have become shallow, green meadows cut into strips by the tracks of caterpillar tractors, half-abandoned villages, farm machinery rusting under the open sky, prefabricated houses, and fields overgrown with weeds all characterize the residents of this or that village, of this or that region no less than an unattractive and neglected apartment tells us something about its inhabitants.

The number of people on the earth continues to grow, as does the amount of technology. Logically, these factors would intensify the need of people for beauty and their very sensitivity to what is beautiful. Take Japan, that most amazing country, the brightest flower of our planet. This populous nation, which lives under incredibly crowded conditions on several pebbles in a vast ocean, and which is uncertain about the geological future of those pebbles, not only has not lost its feeling for what is beautiful, not only has not dulled or deadened its soul, but on the contrary may serve as an example of genuine, profound human civilization. We really should take this as an example, because maybe it represents the future of all other nations, insofar as our entire planet at some point will become just as crowded as it is today on the islands of Japan. But in the process will the other nations preserve that loftiness of spirit? Will they be able to admire the branch of a cherry tree in blossom; to wonder, just as reverently, at the snow-white tent of a mountain; similarly to raise the arranging of a bouquet of flowers to the level of art; to grow, just as patiently, tiny though living gardens that form miniature but genuine landscapes? The feeling for what is beautiful that lives today in the soul of the Japanese people does not represent solely an aesthetic category but also shows courage and heroism. In considering these people we can better understand Dostoevsky's words, that beauty will be the world's salvation.[12]

The well-known Soviet poet Vasily Fyodorov wrote a poem entitled "Prodannaya Venera" (Venus for sale),[13] which is based on the historical fact of the sale of several valuable canvases for the needs of construction and industry. The poem contains the following words: "For the beauty of the coming centuries / We sacrificed beauty."

I understand both the bitterness and the historical optimism of the poet, but still the question arises: Where will the beauty of the coming centuries come from, if we, our generations, do not preserve it in ourselves and do not pass it on to our descendants in an ennobled form and in ever-increasing amounts?

The present as well as future appearance of the earth depends on all of us who live on the earth today; let us do everything we can to keep it beautiful for all time.[14]

NOTES

Vladimir Soloukhin's original notes are identified as such. All other notes are the translator's.

INTRODUCTION

1.　John B. Dunlop, *The Faces of Contemporary Russian Nationalism* (Princeton, N.J.: Princeton University Press, 1983), 34. Dunlop's citation is: "Vladimir Soloukhin, *Rabota*, Moscow, 1966, p. 12."

2.　An excerpt from an unpublished book by Soloukhin, *Posledniaia stupen'* (The last step).

3.　Dunlop, 32.

4.　Ibid., see 32 ff.

5.　Published over the years in various Soviet periodicals and collected in three separate volumes: *Kameshki na ladoni* (Moscow: Sovetskaia Rossiia, 1977), *Kameshki na ladoni* (Moscow: Molodaia gvardiia, 1982), and *Kameshki na ladoni* (Moscow: Sovremennik, 1988).

6.　See Dunlop, 286–90, for his policy recommendations on how the U.S.A. should respond to the Soviet Union/Russia, which include background material on the reasons why Russian nationalists feel that their ethnic group has been treated unfairly by the Soviet Union.

7.　Vladimir Soloukhin, "Chto nas rodnit: zametki pisatelia," *Literaturnaia gazeta*, Feb. 6, 1962, 3. Also quoted in Dunlop, 133.

8.　Dunlop assesses Soloukhin's orientation as that of a "centrist" Russian nationalist in his "The Contemporary Russian Nationalist Spectrum," *Radio Liberty Research Bulletin, Russian Nationalism Today* (Dec. 19, 1988): 6. My own meetings and correspondence with Soloukhin over a seven-year period (1985–92) lead me to conclude that his nationalist orientation is broad-minded, sophisticated, and contradictory: he is a devout Russian Orthodox, a self-proclaimed monarchist, and yet an internationalist thinker inclined to appreciate cultural and religious diversity. I would locate his views closer to those of Likhachev, who is acknowledged as a liberal Russian nationalist.

9.   "Spiritual internationalism was always an intelligent characteristic of the Russian intelligentsia." Dmitrii Likhachev, "Veriu v russkuiu intelligentsiiu," *Literaturnaia gazeta*, Dec. 28, 1988, 4.

10.   Gerald E. Mikkelson, "Vladimir Soloukhin: The Poet Gathering Stones," unpublished paper, 10–11. Mikkelson cites the following lines from Soloukhin's essay "A Time to Gather Stones": "After all, this is our national treasure. And not only ours, but the treasure of the entire, as they say, cultured and progressive human race" [my translation—V.N.].

11.   Kathleen F. Parthé, "Time, Backward! Memory and the Past in Soviet Russian Village Prose," *Kennan Institute Occasional Paper 224* (Washington, D.C.: Kennan Institute for Advanced Russian Studies, Woodrow Wilson Center, 1988), 1. See also Parthé, *Russian Village Prose: The Radiant Past* (Princeton, N.J.: Princeton University Press, 1992), for a sensitive and comprehensive discussion of this rich, yet insufficiently studied, body of literature.

12.   Victor Terras, "Phenomenological Observations on the Aesthetics of Socialist Realism," *Slavic and East European Journal* 23, no. 4 (Winter 1979): 445.

13.   David Lowe, *Russian Writing Since 1953: A Critical Survey* (New York: Ungar, 1987), 92.

14.   Feliks Kuznetsov has called village prose *"liricheskii reportazh"* (lyrical reportage) in "The Fate of the Village in Prose and Criticism," *Soviet Studies in Literature* 10, no. 2 (Spring 1974): 66, translated by William Mandel from Kuznetsov, Sud'by derevni v proze i kritike," *Novyi mir*, no. 6, 1973.

15.   Ibid., 67.

16.   Specific publication information on each individual essay is provided in the Notes. The four essays were published in the 1970s in the journal *Moskva*, an event that caused one of its editors, M. N. Alekseev, to be chastened by M. V. Zimianin, with the following words: "The 110th anniversary of V. I. Lenin's birth is approaching, and you're publishing essays about Optina Monastery and other monasteries!" [my translation—V.N.]. Quoted from "Sobesedniki na pominkakh: Vladimir Soloukhin v besede s korrespondentom *L.G.*, Liana Polukhina," *Literaturnaia gazeta*, Nov. 25, 1992.

17.   *Vremia sobirat' kamni* (Moscow: Sovremennik, 1980).

18.   *Pis'ma iz Russkogo Muzeia* (Moscow: Molodaia gvardiia, 1990), 229–414, and *Vremia sobirat' kamni* (Moscow: Pravda, 1990), 271–504. *Pis'ma iz Russkogo Muzeia* also includes the title collection of essays, and those of *Chernye doski; Vremia sobirat' kamni* contains the essays of the title, as well as *Chernye doski* and the "letters" of *Prodolzhenie vremeni* (The continuation of time), *Nash sovremennik*, no. 1 (January 1982): 18–126. This last work, *The Continuation of Time*, is closely related in subject matter and style to the essays of *A Time to Gather Stones*. When I spoke with Soloukhin in Moscow on July 20, 1991, about his essays, he pointed out that the essays of *The Continuation of Time* are contiguous (*primykaiut*) with those of *A Time to Gather Stones*.

19.   Personal interview, Moscow, July 20, 1991.

20.   "Civilization and Landscape" was republished in the following collection of Soloukhin's essays: *Volshebnaia palochka* (Moscow: Moskovskii rabochii, 1983), 235–45.

21.   The buildings at Shakhmatovo, including the Bloks'-Beketovs' house, are in the process of being restored; on November 17, 1987, the Soviet government returned Optina Monastery to the Russian Orthodox church.

22.   Mikkelson, 14.

23.   *S liricheskikh pozitsii* (Moscow: Sovetskii pisatel', 1965).

24.   For a fuller discussion of Soloukhin's philosophy of art, see ibid., in particular the essays "Poeziia i vremia" (Poetry and time), 17, 32, and "Otvety na voprosy, predlozhennye zhurnalom *Voprosy literatury*" (Answers to questions posed by the journal *Literary Issues*), 67.

25.   "Considering the harshness of Soviet literary politics, Soloukhin's publicistic works of the past fifteen years, especially the cycle called *A Time to Gather Stones*, constitute their own kind of *podvizhnichestvo* worthy of the Optina tradition." Mikkelson, 16.

A VISIT TO ZVANKA

"Poseshchenie Zvanki" first appeared in *Moskva* 7 (1975): 187–98.

1.   Nikolai Roerich (1874–1947): Russian painter whose work was inspired by Scandinavian and medieval Russian culture.

2.   Reference to the territory and Scandinavian people identified with the establishment of the Kievan state in the 9th century A.D.

3.   Verst: a Russian unit of distance equal to 0.6629 miles.

4.   Volkhovstroi: a power station.

5.   The Volkhov front during World War II.

6.   Catherine II (the Great), reigned 1762–96.

7.   "Vel'mozha" (Nobleman), November 1794. Unless otherwise indicated, dates for the works cited in these essays refer to the date of writing; and likewise (unless otherwise indicated) all translations are my own.

8.   "Na ptichku" (To a bird), written by Derzhavin in either 1792 or 1793.

9.   "Na Novyi God" (For New Year's Day), written by Derzhavin in December 1780 or January 1781. In his verse Derzhavin usually referred to his first wife, Ekaterina Yakovlevna Bastidon (1760–94), as "Plenira."

10.   "K pervomu sosedu" (To my first neighbor), 1780.

11.   *Evgenii Onegin*, 1823–31, chap. 1, stanza 16, ll. 7–14.

12.   "Pokhvala sel'skoi zhizni" (In praise of country life), 1798.

13.   From "Excerpts from Onegin's Travels," which were not included in the final version of *Evgenii Onegin*. See A. S. Pushkin, *Sochineniia v trekh tomakh* (Moscow: Khudozhestvennaia literatura, 1978), 3:169.

14.   "K Felitse" (Ode to Felitsa), 1782; "Na smert' Kniazia Meshcherskogo" (On the

death of Prince Meshchersky), 1779.

15. "God," 1784. Hereafter in the text I refer to the poem by its English title.

16. Vasilii Trediakovskii, *Telemakhida*, published in 1766.

17. Peter I (the Great), reigned 1682–1725.

18. Aleksandr Radishchev, *Puteshestvie iz Peterburga v Moskvu* (Journey from St. Petersburg to Moscow), first published in May 1790. "The monster" is a reference to the institution of serfdom. Soloukhin reverses Trediakovskii's original word order of "savage, gigantic" (see *Telemakhida*, vol. 2, bk. 18, verse 514).

19. Trediakovskii, "Parafrazis vtoryia pesni Moiseevy" (Paraphrase of the second song of Moses), 1752.

20. This last line is taken from an English translation of Derzhavin's poem by J. K. Stallybrass, first published in *The Leisure Hour* (London), May 2, 1870. The translated poem appears under the title "Ode to the Deity" in *Anthology of Russian Literature*, in two parts, part 1, ed. Leo Wiener (New York: Benjamin Blom, 1902; reissued 1967), 379–82.

21. Incidentally, I once was dumbfounded as I was leafing through the magazine *Yunyi naturalist*, a subscription to which I had given my daughter. In it I read the following, in black and white: "One can consider it a proven fact that man did not evolve from the monkey, but rather that the development of these two biological branches progressed in parallel, independent directions." [SOLOUKHIN'S NOTE]

22. I later gave this to Anatolii Yelkin, a collector of World War II memorabilia, especially of the naval forces. [SOLOUKHIN'S NOTE]

23. A street in Moscow.

24. Aleksandr Griboedov (1795–1829): Russian playwright best known for his social comedy *Gore ot uma* (Woe from wit), written in 1822–23.

AKSAKOV'S TERRITORY

"Aksakovskie mesta," first published in *Moskva* 8 (1976): 180–201.

1. Soloukhin has his own essay in mind.

2. From the Russian folktale "Kolobok" (The gingerbread man).

3. Young Pioneers: a children's communist organization in the USSR.

4. *Zolotnik* (archaic): Russian measure of weight, equivalent to approximately 4.26 grams. The word is appropriate because it can be translated as "a small gold coin" and is used to refer to something that is small but valuable.

5. From *The Legend of Sleepy Hollow*, by Washington Irving, 1819–20.

6. Moskovskii Gosudarstvennyi Universitet (Moscow State University).

7. *Kisel'*: a kind of starchy jelly.

8. Serezha: diminutive of Sergei.

9. In Russian: *kliuchnitsa*, "keeper of the keys."

10. *Izba*: Russian peasant's log house.

11.   *Kvas:* fermented Russian drink similar to beer, made from rye or barley.

12.   Unidentified poem, most likely dating from the second half of the nineteenth century.

13.   As the saying goes, I'm selling it for the same price as I paid for it. The quotation from Mashinskii is precise. Of course, in working on the essay I couldn't refrain from juxtaposing the two texts, if only out of my own interest. In all likelihood the statement about the textual coincidence is too strong. But in spirit and intonation both texts are actually very much alike. Here, in my opinion, it is necessary to take additional circumstances into account as well: the same manifestation of nature's forces—a snowstorm—is described. And in the same places—the region of Orenburg. *The Captain's Daughter* is seemingly authored by Grinev—i.e., by a person of the eighteenth century—and Pushkin consciously created his *nouvelle* in the style of the eighteenth century, whereas Aksakov in his style was inclined involuntarily toward that era. On the second page of Pushkin's *nouvelle* we encounter the sentence: "But since we served wine only at dinner, and only by the liqueur glass, my Beaupré very soon got used to Russian *nastoika* [a kind of liqueur—V.N.]," etc. This model is not characteristic of Pushkin in his other *nouvelles* that aren't styled after those of the eighteenth century, but here and there we encounter it in Aksakov. So the similarity of the passages juxtaposed by S. Mashinskii could even be accidental. But maybe he's right. The suddenness of a severe snowstorm in steppes that had been quiet and calm immediately preceding it really coincides in both texts. Moreover, in Pushkin's *nouvelle* the landscape and setting played a secondary role, and his descriptions of them are brief. Aksakov, however, takes nine pages to describe a severe snowstorm out of purely scenic considerations, exclusively for the sake of describing a snowstorm. The sketch was noticed even by the critics. *Moskovskii telegraf* especially lavished praise on it: "The masterly depiction of a winter blizzard in the Orenburg steppes. . . . If this is an excerpt from a novel or *nouvelle,* we congratulate the public on [the occasion of—V.N.] a literary work."

Be that as it may, here are both the passages in question by Aksakov and Pushkin:

A white cloud quickly rose and grew from the east, and when the last pale rays of the setting sun had disappeared behind the mountain, an enormous cloud of snow had already covered the greater part of the sky and had begun to produce a fine, snowy dust. . . . The white cloud of snow, as vast as the sky, stretched across the entire horizon, and the last light of the fading red evening glow was quickly covered by a dense mantle. Suddenly night fell. . . . The blizzard came in all its fury, with all of its terrors. . . . Everything flowed together, everything became confused . . . the earth, and air, and sky were transformed into an abyss of surging snowflakes. . . .—Aksakov

A small cloud turned into a white storm cloud that rose with difficulty, grew, and gradually covered the sky. A fine snow began to fall—and suddenly started falling in thick flakes. The wind howled; a blizzard had developed. In a single instant the dark sky merged with a sea of snow. Everything disappeared. "Well, sir," cried the coachman, "we're in trouble: it's a storm!"—Pushkin [SOLOUKHIN'S NOTE]

14. Vladimir Soloukhin, *Kaplia rosy, Znamia* 1 (1960): 54–106, and 2 (1960): 3–76.

15. Mikhail Bubennov, *Belaia bereza* (Moscow: Sovetskii pisatel', 1948).

16. Lermontov, "Rodina," 1841.

17. Vladimir Soloukhin, *Tret'ia okhota, Nauka i zhizn'* 5 (1967): 97–113; 6 (1967): 97–113, and 7 (1967): 97–112.

18. Reference to the All-Russian Society for the Preservation of Historical and Cultural Monuments (Vserossiiskoe Obshchestvo Okhrany Pamiatnikov Istorii i Kul'tury; acronym: VOOPIK), which was established in 1966. See the chapter titled "Voluntary Societies" in Dunlop, *The Faces of Contemporary Russian Nationalism*, 63–92, for further information on these societies.

19. Chertova does not provide the details of the second notice (actually a letter to the editor), which are as follows: V. Savel'zon, "Na beregu Buguruslanki," 6.

20. The Feast of the Sign (*prazdnik znameniia*): one of the holy days marked by the Russian Orthodox Church to honor the sign given to the Virgin Mary that she would become the mother of Jesus Christ, and also to honor one of the icons of Our Lady of the Sign (Ikona Bozh'ei Materi Znameniia). The feast is celebrated on November 27 (O.S.) in Novgorod and on September 8 (O.S.) in Kursk, two cities (of eight) in which originals of the icons of Our Lady of the Sign are located. The church in the hamlet of Znamenskoe in all likelihood was dedicated to one of these icons.

21. "Great Reforms": Major reforms during the reign of Alexander II (1855–81) that included the abolition of serfdom in 1861, as well as reforms in the government, the judiciary, finance, and education.

22. The zemstvo chief, or land captain, had no relationship to the zemstvo form of self-government, but rather was charged with the "direct bureaucratic supervision" of the peasants. See Nicholas V. Riasanovsky, *A History of Russia*, 2nd ed. (New York: Oxford University Press, 1969), 435.

23. Mashinno-Traktornaia Stantsiia (Machine and Tractor Station).

24. *Tovarishch* (comrade); hereafter designated throughout these essays by "T."

25. Not all of them have small holes that have been torn. Some have been recopied in longhand—i.e., copied from copies that were found in folders at my request by Aksakov enthusiasts who were aggrieved over the state of the Aksakov memorial places. [SOLOUKHIN'S NOTE]

26. Remontnaia-Traktornaia Stantsiia (Tractor-Repair Station).

27. The Regional Executive Committee.

28. Upravlenie Kapital'nym Stroitel'stvom (Department of Major Construction Work).

29. In exactly the same way, incidentally (I remember that I wrote about this in *Vladimir Back Roads*), right during the war the trees in the park in the hamlet of Varvarino of Vladimir

Province were all cut down, on the estate linked memorially with Tiutchev's daughter and with Ivan Sergeevich Aksakov—i.e., the son of the main and most important Aksakov. Those who wish to may read the appropriate chapter of *Vladimir Back Roads*. [SOLOUKHIN'S NOTE]

30.　In Russian, *pomoch*: work in exchange for food.

31.　All-Russian Society for the Preservation of Historical and Cultural Monuments.

32.　Sredvolgovodgiprovodkhoz—Gosudarstvennyi institut proektirovaniia vodnogo khoziaistva Srednei Volgi.

33.　VLKSM: Vsesoiuznyi Leninskii Kommunisticheskii Soiuz Molodezhi.

34.　A car from the Gorky Automobile Plant.

35.　Soloukhin uses the feminine form, *kartofelekopalka*.

36.　Lev Tolstoi's family estate.

37.　Mikhailovskoe . . . Spasskoe-Lutovinovo: family estates of nineteenth-century Russian writers.

GREATER SHAKHMATOVO

"Bol'shoe Shakhmatovo," first published in *Moskva* 1 (1979): 196–206, and 2 (1979): 183–200.

1.　"Podnimetsia dom s mezoninom."

2.　Moscow and Leningrad (now St. Petersburg).

3.　The original Russian states "1784," but this is an error, since Dmitrii Mendeleev was born in 1834 and died in 1907.

4.　"Pogruzhalsia ia v more klevera" (1903).

5.　*Volost'*: small rural district.

6.　*Sazhen* = 2.134 meters.

7.　Tipografiia Panteleevykh: St. Petersburg publishing house of the late nineteenth century.

8.　Soloukhin uses the feminine pronoun, most likely because he is contrasting this type of writer with Blok's female relatives.

9.　"Ni sny, ni yav" (March 1921; Blok died in August of that year).

10.　*Zemstvo*: an elective district council in Imperial Russia.

11.　*Aleksandr Blok. Biograficheskii ocherk.*

12.　*Semeinaia khronika*: the book apparently was never published, but a work entitled *Aleksandr Blok i ego mat'*, by M. A. Beketova, was reviewed by V. Piast in *Krasnaia gazeta*, August 7, 1925. This may be *Semeinaia khronika* under a different title or an excerpt taken from it.

13.　*Yat'*: letter of the Russian alphabet in the pre-1917 orthography.

14.　Pood = 36.11 pounds avoirdupois.

15.　A type of mushroom.

16.　"Noch', ulitsa, fonar', apteka" (1912); "Neznakomka" (1906); "Okna vo dvor" (1906);

"Ia prigvozhden k traktirnoi stoike" (1908); "Vechnost' brosila v gorod" (1904); "V kabakakh, v pereulkakh, v izvivakh" (1904); "Na ostrovakh" (1909); "V restorane" (1910). The wine in the last poem is *Ay-Petri*, from the Crimean Mountains.

17. *Dvenadtsat'* (1918).

18. "Skify" (1918).

19. "Na pole Kulikovom" (1908). The Battle on Kulikovo Field took place in 1380: Dmitrii, Prince of Moscow, defeated the Mongols in this decisive, and famous, battle. The battle resulted in a crushing defeat for Mamai, the military leader of the Mongols.

20. "Solov'iny sad" (1914–15).

21. "Osennii den'" (1909).

22. To date, only one volume of Lesnevskii's study has been published: *Put', otkrytyi vzoram; moskovskaia zemlia v zhizni Aleksandra Bloka* (Moscow: Moskovskii rabochii, 1980).

23. "V syrom nochnom tumane" (1912).

24. Evidently buttercups. [SOLOUKHIN'S NOTE] [Blok uses the word *grushovka*, which refers to a type of apple tree; Soloukhin's note explains that Blok actually meant *grushanka*, buttercups.—V.N.]

25. Aleksander Blok, *Avtobiografiia* (June 1915).

26. "Osenniaia volia" (1905).

27. "Osenniaia liubov'" (1907).

28. Aleksandr Aleksandrovich (Blok).

29. *Stikhi o Prekrasnoi Dame* (1901–2); *Nechaiannaia radost'* (1905; collection of poems).

30. *Brichka*: a light carriage, sometimes with a roof.

31. *Raznochintsy*: a group of radicals in Russia of the 1860s (the term translates as "those of diverse social strata"), described by Riasanovsky as "sons of priests who did not follow the calling of their fathers, offspring of petty officials, or individuals from the masses who made their way up through education and effort." See Riasanovsky, *A History of Russia*, 423.

32. Liubov' Dmitrievna (Mendeleeva).

33. Soloukhin misquotes the title, which is "Zapevaiushchii son, zatsvetaiushchii tsvet" (1902); "Vot on—ryad grobovykh stupenei" (1904); "Vechereiushchii sumrak, pover'" (1901).

34. "Zapevaiushchii son, zatsvetaiushchii svet."

35. "Svetlyi son, ty ne obmanesh" (1904; *vechernitsa*: female student).

36. The four poems mentioned are, respectively: "Osenniaia volia," "Starost' mertvaia brodit vokrug," "Devushka pela v tserkovnom khore," and "V lapakh kosmatykh i strashnykh."

37. "Slyshu kolokol. V pole vesna" (1902); "Vstanu ia v utro tumannoe" (1901); "Razletyas' po vsemu nebosklonu" (1914); "On zanesen—sei zhezl zheleznyi" (1914); "Priskakala dikoi step'iu" (1905).

38. E. A. Beketova (1855–92), *Stikhotvoreniia*, posthumous ed. (St. Petersburg: Tipografiia Panteleevikh, 1895).

39. E. A. Beketova, "Siren" (1878).

40. E. A. Beketova, "Solovei" (1887).

41. Lines from unidentified poems, probably written by contemporaries of E. A. Beketova.

42. E. A. Beketova, "Zhuravli" (1888).

43. E. A. Beketova, "Na blednom zolote zakata" (1888).

44. "K Muse" (1912).

45. "Starost' mertvaia brodit vokrug" (1905).

46. "Moi zhestokii s belymi nochami" (1908); "Ty otoshla, i ia v pustyne" (1907); "Zakliatie ognem i mrakom" (1907).

47. "Ia—tvar' drozhashchaia. Luchami" (1902); "Pered sudom" (1915); "Inok" (1907); "Zakliatie ognem i mrakom"; "Ravenna" (1909); "Zakliatie ognem i mrakom."

48. "Priblizhaetsia zvuk. I, pokorna shchemiashchemu zvuku" (1912).

49. *Slovo o polku Igoreve* (The lay of the host of Igor; late twelfth century), the most important Russian medieval work.

50. *Vozmezdie* (1910–11).

51. "Ispoved' iazychnika" (1918; an autobiographical story).

52. "Rus" (1906).

53. "Nad ozerom" (1907).

54. "Est' v dikoi roshche u ovraga" (1898); "Ia shel k blazhenstvu. Put' blestel" (1899); "Belyi kon' chut' stupaet ustaloi nogoi" (1905); "Pliaski osennie" (1905); "Na nebe zarevo. Glukhaia noch' mertva" (1900); "Ia voskhodil na vse vershiny" (1904).

55. "Ishchu ognei—ognei poputnykh" (1906).

56. "V syrom nochnom tumane."

57. "Dmitrii Ivanovich played a very large role in the Beketov family, and my grandfather and grandmother were on friendly terms with them. Soon after the liberation of the peasants, Mendeleev and my grandfather went to Moscow Province and purchased two estates near each other in Klinsk District; Mendeleev's Boblovo is located seven versts from Shakhmatovo—I visited it in my childhood, and in my adolescence began to go there often. The eldest daughter of Dmitrii Ivanovich Mendeleev by his second marriage—Liubov' Dmitrievna—became my fiancée. In 1903 she and I were married in the church of the hamlet of Tarakanovo, which is located between Shakhmatovo and Boblovo." (Blok, *Avtobiografiia*). [SOLOUKHIN'S NOTE]

58. Liubochka and Sashura are diminutives for Liubov' and Aleksandr, respectively.

59. "Slovoopolkuigorevskii."

60. *Rodina* (1907–16) includes "Na pole Kulikovom."

61. "Vstrechnoi" (1908); "O doblestiakh, o podvigakh, o slave" (1908; from *Vozmezdie*); "Tak okrylenno, tak napevno" (1906); "Idut chasy, i dni, i gody" (1910); "Ia umer. Ia pal ot rany" (1903); "Fioletovyi zapad gnetet" (1904); "Ia zhivu v glubokom pokoe" (1904); "Mne bitva serdtse veselit" (1901); "Na snezhnom kostre" (1907); "Nasmeshnitsa" (1904); "Poedinok" (1904); "Vliublennost'" (1905); "Pora vernut'sia k prezhnei bitve" (1900); "Ia—mech, zaostrennyis obeikh storon" (1903); "Pogruzhalsia ia v more klevera"; "Syn i mat" (1906);

"Tak okrylenno, tak napevno"; "Dali slepy, dni bezgnevny" (1904); "Golosa skripok" (1910); "Za kholmom otzveneli uprugie laty" (1907); "Syn i mat"; "Tak okrylenno, tak napevno"; "Na pole Kulikovom."

62.　The time of Catherine II (the Great).

63.　"Listok" (1841); "Vykhozhu odin ia na dorogu" (1841).

64.　"Na pole Kulikovom."

65.　Warm felt boots, usually waterproof.

66.　*Khorosho!* (1927).

67.　Poet and friend of Blok.

68.　N. Pavlóvich, "Vospominaniia ob Aleksandre Bloke," *Prometei* (Moscow: Molodaia gvardiia, 1977), 11:231. [SOLOUKHIN'S NOTE]

69.　S. Alianskii, *Vstrechi s Aleksandrom Blokom* (Moscow: Detskaia literatura, 1969), 128. [SOLOUKHIN'S NOTE]

70.　"Pamiati Leonida Andreeva" (In memory of Leonid Andreev; 1919).

71.　P. A. Zhurov, "Sad Bloka," manuscript, Pushkinskii muzei, Moscow. [SOLOUKHIN'S NOTE]

72.　A. Blok, "Pamiati Leonida Andreeva," *Sobranie sochinenii* (Leningrad: Sovetskii pisatel', 1932), 9:191. [SOLOUKHIN'S NOTE]

73.　Soloukhin does not provide the source of the quoted material, but it is most likely the same as that cited in note 72.

74.　The astronomical figures show the hyperinflation that developed after the Revolution of 1917 and ensuing civil war. According to Riasanovsky, "The exchange rate of an American dollar, which had been two rubles in 1914, rose to 1,200 rubles in 1920" (*A History of Russia*, 541). Bank notes were issued in denominations of millions of rubles.

75.　From Blok's drafts for the continuation of the second chapter of *Retribution*, January 1921.

76.　*Subbotnik*: day of voluntary, unpaid labor.

77.　The Russian word *zhivoi* can be translated as either "living" or "lively."

78.　As we learned not long ago, this barn has been demolished as well.

79.　Soloukhin paraphrases two lines from Blok's draft for the continuation of the second chapter of *Retribution*, January 1921: 'I vsei vesennei krasotoiu / Siiaet russkaia zemlia."

A TIME TO GATHER STONES

"Vremia sobirat' kamni" was first published in *Moskva*, no. 2 (1980), 186–212. ("For everything there is a season . . . a time to cast away stones, and a time to gather stones together. . . . "—Eccles. 3:1–5)

1.　A Russian ball game.

2.　Well-known Russian prerevolutionary china factories.

3.　After this essay was published in a journal, the author received many letters from

Ukraine, whose writers informed him that there are actually several museums devoted to G. Skovoroda. One of these letters ends with the sentence "This is how we honor the memory of a great fellow countryman in Ukraine." Here is an instance in which it is heartening, and even pleasurable, to admit my mistake. [SOLOUKHIN'S NOTE]

4. Tolstoi's family estate.

5. Nikolai Nekrasov, "Vlas," published in 1854.

6. Khan Batu, a grandson of Jenghiz Khan, led the Mongol invasion of Europe in the early thirteenth century.

7. One church remains in Kozel'sk, while the island no longer exists. [SOLOUKHIN'S NOTE]

8. The word *prikliuchenie* in those days did not have its contemporary meaning ("adventure"), for if it had it would not have appeared in such a ceremonial document. [SOLOUKHIN'S NOTE]

9. N. N. Anisimov and V. N. Sorokin, *Kozel'sk* (Tula: Priokskoe kniznoe izdatel'stvo, 1967), 22. [SOLOUKHIN'S NOTE]

10. Boris Godunov reigned (1598–1605) during one of Russia's most complex historical periods, the early part of the "Time of Troubles" (which spanned the years 1598–1613).

11. "Krest'ianskikh i bobyl'skikh dvorov net." *Bobyl'*: poor landless peasant.

12. *Obrok*: quitrent (a rent paid by a freeman in lieu of services required of him by feudal custom).

13. Peter I (the Great).

14. In Russian, *stol'niki*: in seventeenth-century Russia *stol'niki* (ranking below the boyars) were responsible for overseeing the diet of the tsar and his family.

15. Leonid (Archimandrite) Kavelin, *Istoricheskoe opisanie Kozel'skoi Vvedenskoi Optinoi Pustyni* (St. Petersburg: V tipografii Morskogo kadetnogo korpusa, 1847).

16. Of course, one should consistently write "so-called church traditions," "so-called spiritual enlightenment," etc. But I hope that contemporary and really enlightened readers will insert the helpful words themselves where necessary. [SOLOUKHIN'S NOTE]

17. *Skit* (English: skete): a small, secluded monastery.

18. *Brat'ia Karamazovy* (The brothers Karamazov, 1881).

19. "A prophet is not without honor, save in his own country, and in his own house" (Matt. 13:57).

20. A black cap (similar to a fez) worn by Russian Orthodox monks and clergy.

21. Dmitrii Zhukov, "Kto vosstanovit pamiatnik," *Literaturnaia Rossiia*, Jan. 20, 1978. [SOLOUKHIN'S NOTE]

22. F. M. Dostoevskii, *Brat'ia Karamazovy*, 1881. [SOLOUKHIN'S NOTE]

23. Unidentified poem by Nikolai Yazykov (1803–47).

24. See Soloukhin's note 27.

25. A. G. Lushnikov, *I. V. Kireevskii. Ocherk zhizni i religiozno-filosofskogo mirovozzreniia* (Kazan': n.p., 1913). [SOLOUKHIN'S NOTE]

26. Ibid. [SOLOUKHIN'S NOTE]

27. Quoted from D. I. Pisarev, "Russkii Don-Kikhot" (Moscow: Gos. Izd. Khud. Lit.), 1:323. [SOLOUKHIN'S NOTE] [Since Soloukhin's note does not include the title of the book or the date of its publication, I offer the following reference for Pisarev's article, which was written in 1862: D. I. Pisarev, *Sochineniia Pisareva*, 6 vols. (St. Petersburg: V tipografii Yu. N. Erlich, 1904–5), 2:217–36.]

28. N. P. Barsukov, *Zhizn' i trudy v Pogodina*, 22 vols. (St. Petersburg: Tipografiia M. M. Stasiulevicha, 1891), 9:36, bk. 5, pp. 470–71. [SOLOUKHIN'S NOTE]

29. P. O. Kulish, *Zapiski iz zhizni Gogolia*, 2 vols. (St. Petersburg: Tipografiia A. Yakobsona, 1856).

30. The absence of commas in this sentence indicates that here "apparently" is not a parenthetical word but rather is used in the meaning of "visibly." [SOLOUKHIN'S NOTE]

31. Cf., for example, F. Dostoevskii's *Brothers Karamazov*, especially the chapter entitled "They Came to the Monastery." [SOLOUKHIN'S NOTE]

32. Isaak Sirin: an ascetic of the Orthodox church (elevated to sainthood) of the eighth century, who wrote many works on religio-philosophical subjects.

33. *Mertvye Dushi*, vol 1 (1842).

34. P. A. Matveev, "Gogol' v Optinoi Pustyni," *Russkaia starina* (February 1913): 303. [SOLOUKHIN'S NOTE]

35. "Optinskaia pustyn' i palomnichestvo v nee russkikh pisatelei."

36. Gogol''s estate.

37. *Vestnik Evropy* 12 (1905): 710. [SOLOUKHIN'S NOTE]

38. Petrashevtsy: a radical group of men, active from 1845 to 1849, whose program included French utopian socialism and elements of political protest.

39. E. Nikolaev, *Po kaluzhskoi zemle* (Moscow: Iskusstvo, 1970), 117. [SOLOUKHIN'S NOTE]

40. Fedor Kuz'mich: traditionally considered to be Tsar Alexander I (1801–25).

41. . . . it was too late to back out (as the Russian saying goes).

42. I. L. Tolstoi, *Moi vospominaniia* (Moscow: Gos. Izd. Khud. Lit., 1969), 254. [SOLOUKHIN'S NOTE]

43. D. P. Makovetskii, "Poslednie dni Tolstogo," *Novyi mir* 8 (1978): 164. [SOLOUKHIN'S NOTE]

44. SPTU—Sel'skoe proizvodstvenno-tekhnicheskoe uchilishche (rural industrial-technical school).

45. A slight exaggeration. People from all over the world will not travel to a place that is not included in a tourists' itinerary. But it's true that they come from all corners of our country. [SOLOUKHIN'S NOTE]

46. The in-laws of Aleksandr Pushkin.

47. N. Voronin, "Oborvannoe puteshestvie," introduction to *Po kaluzhskoi zemle*. [SOLOUKHIN'S NOTE]

48. A. P. Rogotchenko, *Umanskoe chudo* (Kiev: Reklama, 1977). [SOLOUKHIN'S NOTE]

49. *Osvobozhdenie Tolstogo*, 1937.

50. Soloukhin misquotes the line from Akhmatova's *Severnye elegii* (Northern elegies), 1, which was written in September 1940 (Leningrad) and October 1943 (Tashkent), and which reads "(A v Optinoi mne bol'she ne byvat' . . .)." The line translates as "(And I won't go to Optina anymore . . .)."

51. Aleksei Apukhtin, "God v monastyre," 1882–83.

52. Romance: a musical genre; short, lyrical songs popular in nineteenth- and twentieth-century Russia/USSR.

53. The poems by Apukhtin are, respectively, "Para gnedykh" (the 1870s), "Sumasshed-shii" (1890), and "Sumasshedshii." In this last poem Soloukhin misquotes the following lines:

> Da, vasil'ki, vasil'ki . . .
> Mnogo mel'kalo ikh v pole . . .
>
> Vse vasil'ki, vasil'ki,
> Krasnye, zheltye vsiudu . . .
>
> (Yes, cornflowers, cornflowers, . . .
> How many of them gleamed in the field . . .
>
> Only cornflowers, cornflowers,
> Red ones and yellow ones all around . . .)

54. Apukhtin, "Sovremennym Vitiiam" (1861).

55. "God v monastyre."

56. A. Blok, "Bez bozhestva, bez vdokhnovenya," 1921. [SOLOUKHIN'S NOTE]

57. Soloukhin jokes that a writer might be "issued milk" in order to soothe an upset stomach.

58. Pioneer: Young Pioneers, a children's Communist organization in the USSR.

59. Incidentally, someone (one of my acquaintances) who asked me not to mention his name related that he had taken part in the removal of the cross from the Cathedral of the Entrance of the Holy Virgin into the Temple when there was still a rest home at Optina. In his words: "The physical education instructor, Pet'ka Sokolov (his real name), came up to me and said that he had been charged with removing the cross from the cathedral and that I should help him. Well, I agreed. Because of my youth. And I must say that during the process I experienced one of the most terrifying moments of my life. While we were unscrewing the screws we held onto the cross, but when it flew to the earth we no longer had anything to hold onto, and I thought that my hair would turn gray." [SOLOUKHIN'S NOTE]

60. Fishing "na sezhu": a local, ancient method of catching fish.

61. Validol: a common medicine in Russia that is used as a tranquilizer.

62. Nauchno-issledovatel'skii institut: Scientific-Research Institute.

63. Alexander III: reigned 1881–94.

64. Lyovochka: diminutive of "Lev."

65. Gos. Lit. Muzei, *Letopis'. L. N. Tolstoi* (1938), bk. 2, pp. 314–15. [SOLOUKHIN'S NOTE]

66. Makovetskii, 165. [SOLOUKHIN'S NOTE]

67. Arshin (archaic) = 0.71 meters.

68. In Moscow, firms such as "Melodiia," "Mul'tfil'm," and "Diafil'm" are housed in former churches. It's terribly uncomfortable for them there: I was there and saw this myself. [SOLOUKHIN'S NOTE]

69. Old Believers cross themselves with two fingers, while the Orthodox use three.

70. In the original: *kleikie listy* (a phrase associated with *The Brothers Karamazov*, pt. 2, bk. 5, chap. 3: Ivan tells Alesha that he loves the "sticky green leaves as they open in spring . . . ").

CIVILIZATION AND LANDSCAPE

A paper presented by Vladimir Soloukhin at the Nobel symposium "The Feeling for Nature and the Landscape of Man" held in Göteborg, Sweden, on Sept. 10–12, 1978. (Soloukhin's presentation took place on Sept. 12.) The paper was translated from the Russian original into Swedish by Bengt Eriksson, and subsequently from this Swedish text into English by Wesley Sykes (commissioned by the Royal Society of Arts and Sciences of Göteborg) for the published proceedings of the symposium, *Nobel Symposium 45: The Feeling for Nature and the Landscape of Man*, ed. Paul Hallberg (Göteborg: Royal Society of Arts and Sciences, 1980), 107–20.

1. *Entsiklopedicheskii slovar'*, 82 vols. (St. Petersburg: F. A. Brokgauz i I. A. Efron, 1890–1904); four supplementary volumes were issued by the same publisher during the years 1905–7.

2. Jean Zeitoun, "La Notion de paysage," *L'Architecture d'aujourd'hui* 145 (September 1969): 30. [SOLOUKHIN'S NOTE]

3. Reference to the military campaign mounted by Alexander II to liberate the Bulgarians from Turkish oppression. Russia declared war on Turkey in April 1877.

4. Yves Betolaud, "Urbanisation et nature," *L'Architecture d'aujourd'hui* 145 (September 1969): 45–47. [SOLOUKHIN'S NOTE]

5. L. N. Tolstoi, *Sobranie sochinenii*, 14 vols. (Moscow: Gos. Izd. Khud. Lit., 1951), 3: 157–58. [SOLOUKHIN'S NOTE]

6. From Lermontov's poem "Sosed" (Neighbor), written in February 1837, first published in 1840.

7. A. T. Yagodovskaia. *O peizazhe* (Moscow: Sovetskii khudozhnik, 1963). [SOLOUKHIN'S NOTE]

8. Red-figured vase painting influenced Greek and Roman painters in the last quarter of the sixth century B.C. Instead of following traditional formulas, this type of painting emphasized the direct observation of nature and fluidity of movement.

9.   This sentence may introduce the second reason, which is not made explicit by Soloukhin.

10.   Moscow: Gosstroiizdat, 1960.

11.   Alexander Solzhenitsyn has also noted the impact Russia's churches have had on her landscape—and on the human spirit: "Traveling along country roads in central Russia, you begin to understand why the Russian countryside has such a soothing effect. It is because of its churches. They rise over ridge and hillside, descending toward wide rivers like red and white princesses, towering above the thatch and wooden huts of everyday life with their slender, carved and fretted belfries. From far away they greet each other; from distant, unseen villages they rise toward the same sky. Wherever you may wander, over field or pasture, many miles from any homestead, you are never alone . . . the dome of a belfry is always beckoning to you. . . . " From Solzhenitsyn's prose poem "Puteshestvie vdol' Oki" (A journey along the Oka), first published in Germany under the title *Im Interesse der Sache* (West Berlin: Luchterhand Verlag, 1970). The English citation is taken from Alexander Solzhenitsyn, *Stories and Prose Poems*, trans. Michael Glenny (New York: Farrar, Strauss, and Giroux, 1971).

12.   "Ladies and gentlemen," he [Ippolit—V. N.] cried in a loud voice to everyone present, "the prince maintains that beauty will save the world. . . . What kind of beauty will save the world?" *The Idiot* (1868), pt. 3, chap. 5.

13.   V. Fedorov (1918–84), "Prodannaia Venera," 1956.

14.   Soloukhin concludes his address with words well-known in the Russian Orthodox Church's liturgy: "vo veki vekov" (usually translated as "to the ages of ages").